Palgrave Studies in Sound

Series Editor
Mark Grimshaw-Aagaard
Musik
Aalborg University
Aalborg, Denmark

Palgrave Studies in Sound is an interdisciplinary series devoted to the topic of sound with each volume framing and focusing on sound as it is conceptualized in a specific context or field. In its broad reach, Studies in Sound aims to illuminate not only the diversity and complexity of our understanding and experience of sound but also the myriad ways in which sound is conceptualized and utilized in diverse domains. The series is edited by Mark Grimshaw-Aagaard, The Obel Professor of Music at Aalborg University, and is curated by members of the university's Music and Sound Knowledge Group.

More information about this series at
http://www.palgrave.com/gp/series/15081

Ryoko Sasamoto

Onomatopoeia and Relevance

Communication of Impressions via Sound

palgrave
macmillan

Ryoko Sasamoto
SALIS
Dublin City University SALIS
Dublin, Ireland

Palgrave Studies in Sound
ISBN 978-3-030-26317-1 ISBN 978-3-030-26318-8 (eBook)
https://doi.org/10.1007/978-3-030-26318-8

This Palgrave Macmillan imprint is published by the registered company Springer Nature Switzerland AG.
The registered company address is: Gewerbestrasse 11, 6330 Cham, Switzerland

For my parents who supported me throughout my life, and for Diane, who taught me how to do research.

Acknowledgements

Prof. Diane Blakemore is the best supervisor one could ever wish for. While this book is not based on my PhD thesis, I couldn't have written this without having worked under her guidance. I still remember when I first met her. She gave me three extremely philosophical papers. I am certain that it was her way of assessing what I was capable of and how I work. Within a week or so of supervising me, she identified my strength—an eye for data with a bit of theory (in a tongue-in-cheek way!). That's what I still do—my research is firmly founded in theory, relevance theory in particular, while I have a very strong applied aspect with real data. If Diane hadn't identified my style of working then, I might not be doing this now. So, thank you, Diane!

Another very special person who inspired me is Olivia Rohan. Thank you, Olivia, for such inspirational discussions on onomatopoeia. I hope this book does you justice.

I have also had the privilege to meet the finest scholars in relevance theory. I was particularly privileged to work with Prof. Deirdre Wilson, who provided extremely insightful comments on earlier stages of my work on onomatopoeia.

I would also like to thank Tim Wharton for inspiring me to work on onomatopoeia and other aspects of meaning beyond verbal communication.

Kate Scott is another inspiration—throughout the process of writing this book, Kate and I encouraged each other and remotely worked together. Kate is an inspiration, a good friend, and a sounding board in one—one cannot wish for more.

I would also like to express my gratitude to friends and colleagues in SALIS—past and present. In particular, I would like to thank three colleagues I always look up to: Aileen Pearson-Evans, Dorothy Kenny, and Minako O'Hagan. I hope that one day, I can support colleagues in the way you support me.

I should not forget how I started out as a Linguistics student. I wish to thank Prof. Seiji Uchida, Prof. Akiko Yoshimura, and Prof. Ayumi Suga of Nara Women's University, for guiding me thorough my early years in research. I would also like to thank friends who supported me throughout my student days, in particular, Kazuyo Murata, Rebecca Jackson, and Adam Gargani.

I wish to express my sincere gratitude to the publishers and manga-artists who gave permission for their copyright material to be included in this book, including Asumi Yoshino and Gekkan Korokoro Comic; Cookpad; Kani Doraku; Mieko Osaka and Hiyoko Club; ONE/Yusuke Murata and Shueisha; Rin Mikimoto and Kodansha; Usao Hirara and Neko Publishing; Yuto Tsukuda/Shun Saeki and Shueisha. I would also express my sincere gratitude to Cookpad Inc. and Dr. Jun Harashima of Cookpad Inc. for providing me with the user generated content parallel corpus.

This book is supported under the Faculty of Humanities and Social Sciences Book Publication Scheme, Dublin City University.

Last, but not least, I would like to thank Phil and Alfie for their support and patience. You keep me going.

Contents

List of Figures

List of Tables

1

Introduction

Onomatopoeia is often defined as words that mimic sounds. According to the *Oxford English Dictionary* (*OED*), it refers to "the formation of a word from a sound associated with what is named". Standard English examples include *buzz*, *meow*, *crash* and *splash*. Onomatopoeia presents an interesting challenge to the assumption that the link between word form and meaning is completely arbitrary (de Saussure 1959), since the sounds of onomatopoeic words seem to resemble or imitate (at least part of) their interpretations. For example, there is something about the word *buzz* that resembles the sound made by a bee. Some scholars, therefore, argue that onomatopoeia poses a specific challenge for de Saussure's notion of language arbitrariness.

Different terminologies are used to categorise words that fall within this description, such as mimetics, expressives, and ideophones (cf. Dingemanse and Akita 2016, 2019). In Japanese linguistics, the term *mimetics* is generally used rather than *onomatopoeia*. Another term that is often used is *ideophone*, which, according to Akita and Dingemanse (2019), supersedes what is generally covered as *onomatopoeia* and includes a wide range of words that denote imagery from different sensory domains such as motion, texture, states, and sounds. While such distinctions could

R. Sasamoto, *Onomatopoeia and Relevance*, Palgrave Studies in Sound,
https://doi.org/10.1007/978-3-030-26318-8_1

be useful in a descriptive study of *onomatopoeia*, it is not within the scope of this study to make such a distinction and, as such, the term *onomatopoeia* will be used to cover them all. Note that the use of the term *onomatopoeia* in this study is for convenience and is not to be taken as having any particular commitment to the debate on terminology.

According to Hinton et al. (1995, 10–11), the functions of onomatopoeia include communication of mimicry, expression of internal states, expression of social relationship (diminutives, vocatives), salient characteristics of objects and activities (movement, shape, etc.), and grammatical and discourse indicators (intonation markers, distinction between parts of speech, expression of evaluation and affect towards objects).

While onomatopoeia can be used across different genres, especially in languages with ample onomatopoeia such as Japanese, it is particularly prevalent in literary work and other sensual discourse such as food writing, as demonstrated in examples (1) to (3):

(1) How they *tinkle, tinkle, tinkle*,[…] /From the bells, bells, bells, bells, /Bells, bells, bells— / From the *jingling* and the *tinkling* of the bells.
(Edgar Allan Poe, "The Bells", my italics)

(2) there is something so welcoming about a big bowlful, the rich, smooth, eggy cream waiting to *ooze* out on the spoon that breaks through the tortoiseshell disc on top.
(Nigella Lawson, "Crème Brulée", 2001, my italics)

(3) It *SHUSHES* / It *hushes* / The loudness in the road. / It *flitter-twitters*, / And laughs away from me. / It laughs a lovely whiteness, / And whitely *whirls* away, / To be / Some otherwhere, / Still white as milk or shirts, / So beautiful it hurts.
(Gwendolyn Brooks, "Cynthia in the Snow", my italics)

(4) *wakuwaku* shite konsāto ni itta.
Wakuwaku-do concert to went.
"I was excited to go to the concert."

In (1), the onomatopoeic word *tinkle* is used to describe the sound of bells, and *over sprinkle* and *twinkle* to describe the manner in which a star flickers. It also contains *tinkling* and *jingling*, which again describe the sound of bells. Examples (2) and (3), like *twinkle* in (1), contain the use of phenomimes (*gitaigo*), where onomatopoeia is used to describe non-auditory-based senses. In (2) the onomatopoeic expression *ooze out* describes the way in which cream is about to come out onto the spoon, while *hushes* and *shushes* in (3) describe the manner in which snow silences the world. In (4), the gijogo (psychomime) *wakuwaku* describes the person's excited emotional state.

It is generally accepted that onomatopoeia/mimetics evoke some kind of affect and feeling. For example, Kita argues that:

> Japanese mimetics are a class of words that are not only referential but also evoke a vivid at-the-scene feeling [...] In most cases, a mimetic evokes some complex combination of sensory inputs and affect, which can be described more accurately as impression than as sensation.
>
> (Kita 1997, 381)

He further argues that "Japanese mimetics have a unique psychological effect. They evoke vivid 'images' of an experience, full of affect. This imagery is not only visual but can also be based on other perceptual modalities and physiological states" (Kita 1997, 386).

However, it is not entirely clear what onomatopoeia encodes or how it evokes such feelings. Nor is it clear how such feelings are related to what onomatopoeia encodes. This is not to deny what was observed in Kita's (1997) analysis of mimetics. Nevertheless, further investigation into the meaning of onomatopoeia is necessary, in order to establish what is linguistically encoded and how such encoded meaning contributes to the communication of feelings.

Similarly, the highly expressive nature of the meaning of onomatopoeia has led scholars to consider that its meaning is somewhat vague and difficult to define. Furthermore, the meanings of onomatopoeic expressions are often ambiguous, and some scholars treat them as polysemous. For example, Inoue (2013) shows how an onomatopoeic expression *gatagata* could mean (i) a specific sound accompanying a specific aspect of an object/state, as illustrated in (5a); (ii) the way in which something is not regular or orderly, as described in (5b); (iii) an uneven and unsmooth texture, as illustrated in (5c) and (5d); or (iv) the state of something that is not running smoothly or successfully, as shown in (5e):

(5) a. mado o *gatagata* to yusutte akeru.
Window ACC ONO QUO shake open
"[subject] shakes the window to open."
b. *gatagata* suru isu ni suwaru.
ONO do chair LOC sit
"To sit on a rickety chair."
c. hanarabi ga *gatagata* ni natta.
Row of teeth SUB ONO to became
"The row of teeth became unstraight."
d. tsume no hyomen ga *gatagata* da.
Nail GEN surface SUB ONO copula
"The surface of the nail is rough."
e. kono jiken de kaisha ga *gatagata* ni natta.
This case with company SUB ONO to became
"This case rattled the company."

As illustrated in example (5), a single onomatopoeia can communicate a range of different meanings. As such, it is extremely difficult to pin down what onomatopoeia encodes. The question is whether onomatopoeia is in fact polysemous, as is often claimed, or whether multiple interpretations of onomatopoeia are a result of the inference process. The main aim of this book is to answer the question: What is the role of onomatopoeia in communication and how do we communicate with onomatopoeia?

1 Studies on Onomatopoeia

As noted by Akita and Tsujimura (2016) and Akita and Dingemanse (2019), the linguistic significance of onomatopoeia (or *mimetics* or *ideophones* in their terminology) has been discussed across languages. In languages where onomatopoeia has a relatively small role to play, such as European languages, it has largely been seen as insignificant for linguistic research. By contrast, in languages where onomatopoeia is highly prevalent, such as Japanese, Korean, and other Asian and African languages, it has often attracted as much scholarly attention as any other class of linguistic form (see, for example, Childs 1995; Hinton et al. 1995; Sohn

1994; Nuckolls 1996; Kakehi et al. 1996; Hamano 1998; Voeltz and Kilian-Hatz 2001; Ibarretxe-Antunano 2006; Dingemanse 2011a, b, c). Indeed, in these languages, onomatopoeia has a much wider morphological and phonological variation than anticipated, as well as syntactic functions.

As observed by Akita and Dingemanse (2019), focus on onomatopoeia (or ideophone in their terminology) studies can be divided into three categories: definition, integration, and iconicity.[1] Let us see how each aspect of onomatopoeia has been discussed.

First, there is an issue of definition, or what "onomatopoeia/ideophone/mimetics" is. The term ideophone was first introduced by Doke (1935), who defined it as follows:

> A vivid representation of an idea in sound. A word, often onomatopoeic, which describes a predicate, qualificative or adverb in respect to manner, colour, sound, smell, action, state or intensity. … It must be pointed out that generally the special rules of length, tone and stress, applicable in ordinary grammatical forms, differ considerably in the case of ideophones.
> (Doke 1935, 118–119)

This "vividness" has since been central to the definition of onomatopoeia. For example, Kita (1997) defines Japanese mimetics as:

> a class of words that are not only referential but also evoke a vivid at-the-scene feeling. […] In most cases, a mimetic evokes some complex combination of sensory inputs and affect, which can be described more accurately as impression than as sensation.
> (Kita 1997, 381)

Following on from Doke's (1935) definition, Dingemanse (2012) defines ideophones as "marked words that depict sensory imagery". In both Kita's (1997) and Dingemanse's (2012) definitions, it is assumed

[1] Another aspect of studies on onomatopoeia is the relationship to language acquisition (Yoshida 2004; Yoshida and Smith 2003; Maurer et al. 2006; Parault and Schwanenflugel 2006; Haryu and Zhao 2007; Imai et al. 2008; Nygaard et al. 2009; Wrembel 2010; Kantartzis et al. 2011; Laing 2014, 2017, 2018, 2019).

that they are treated as a distinct group of words. Kita (1997) explicitly treats them as "a class of word" (Kita 1997), while Dingemanse (2012) does so less explicitly. However, the implication of this current study will be that onomatopoeia is a property, rather than a word class, as I will show throughout this book.

Another aspect of definition discussed by Akita and Dingemanse (2019) is related to semiotic characteristics of onomatopoeia. Drawing on the semiotic distinction between *depiction* and *description*, they argue that ideophones fit in the category of *depictions*, which are characterised by virtue of iconicity (similarity between form and meaning), as opposed to *descriptions*, which are characterised by "arbitrary form-meaning mappings", that seem to suggest the verbalness. Their distinction between depictions and descriptions is somewhat parallel to the distinction between the analytic dimension and the affecto-imagistic dimension first proposed by Diffloth (1972) and adopted by Kita (1997). While making such distinctions captures intuitions about onomatopoeia and its meaning, it is not clear if such a binary relation allows for cases of highly lexicalised onomatopoeia that have established lexical meaning, as well as vividness/imagery. I will discuss this further in Chap. 4.

Another aspect of onomatopoeia that has been attracting scholarly attention is its phonological, morphological, and morphosyntactic integration. Many scholars (such as Asano 1978; Atoda and Hoshino 1995; Ono 2007; Kakehi et al. 1996; Hinton et al. 1995; Zwicky and Pullum 1987; Hamano 1998) focus on the morphophonological and syntactic characteristics of onomatopoeia to find regularities in its idiosyncratic properties. Such regularities then led scholars' interest to explore the generalisability of regularities found in onomatopoeia and to seek whether this word class might shed light on aspects of human language itself (Akita and Tsujimura 2016, 133–135). Still, as Akita and Tsujimura (2016) state, such generalisability of regularities found in this word class of onomatopoeia has not yet been fully discussed, and the research focus is still on the description of characteristics of particular (sets of) onomatopoeia.

Yet, there are certain linguistic characteristics of ideophones that have emerged which suggest that ideophones do have prototypical characteristics of being a word class, while exhibiting some deviation from major lexical categories such as verbs and nouns. Phonologically speaking, ideophones involve a marked use of speech sounds that do not appear in native phonological systems. For example, while the consonant /p/ rarely appears in the Japanese native system, there are a number of onomatopoeic expressions that use the sound. Akita and Dingemanse (2019) also report similar cases in Iroquoian languages (Mithun 1982) and Korean (Kwon 2018). McCawley (1968) presented similar phonological categories for Japanese, which included Yamatokotoba, mimetics, Sino-Japanese, and loan words. In terms of morphological and morphosyntactic integration of ideophones, it has been well documented that while onomatopoeia rarely involves regular morphological processes,[2] it often goes through word formation such as reduplication or lengthening.

Finally, another aspect that is often discussed is the lexicality and iconicity (or mimeticity) of onomatopoeia. Iconicity is often discussed in terms of *sound symbolism*, which has dominated research on onomatopoeia. As the sound-symbolism approach is assumed by many scholars who work in the area, either explicitly or implicitly, it is worth examining details of this approach. I will come back to this in Chap. 2. Meanwhile, as iconicity and lexicality are often considered to be in inverse relation, it is worth noting what has been discussed in terms of lexicality.

It is generally acknowledged, often implicitly, that onomatopoeia is located on the continuum of a firmly lexicalised cluster to a novel cluster. As I will show in Chaps. 3 and 4, some researchers assign degrees of "lexicality" (or lexicalisation) to onomatopoeia, which can be determined based on the possibility of whether they can be inflected and whether they can occur in a quotation. For example, Kadooka (2005) illustrates degrees of lexicality ranging from the least lexicalised expressions, such as English *hjckrrh* (an exclamation used by Lewis Carroll in *Alice in Wonderland*) and Japanese *bakyuun* (sound of firing a gun), through middle cases, such as *meow* and *pop* in English and *karari* (onomatopoeia for dryness) and *sowasowa-suru* (nervous) in Japanese, to the most lexicalised cases such as English *chatter* and Japanese *odoroku* (surprised). Flyxe (2002) also lists expressions such as *kippari* (clearly) and *odoroku* as

[2] As I will show in Chap. 4, in some highly lexicalised cases onomatopoeia can go through a regular semantic composition process.

examples of highly lexicalised onomatopoeia, and expressions such as *zabun* (splash) or *gān* (an expression often used when someone is shocked) as among the least lexicalised examples of onomatopoeia. Similarly, Flyxe (2002) describes the degree of lexicalisation in terms of semantics and syntactic features.[3]

Observations made in such studies again capture intuitions about differences in the level of wordness of onomatopoeia. Indeed, as Hamano (1998) argues, certain onomatopoeic expressions are so highly lexicalised that some are no longer seen as onomatopoeic. Such expressions include *korogaru* (to roll), which originated from an onomatopoeia *korokoro*, and *isogu* (to hurry), which originated from an onomatopoeic expression *isoiso*. The question here is what it means for *communication* if an expression is highly lexicalised or novel. In this book, I will draw on a relevance-theoretic notion of the *showing–saying* continuum and argue that highly lexicalised onomatopoeia has a strong aspect of the *saying* element, while less established onomatopoeia might involve a lesser *saying* element, providing little coded evidence for communication. This will be discussed in detail in Chaps. 3 and 4.

As I have just shown, research into onomatopoeia has been mostly concerned with the linguistic characteristics of onomatopoeia, including phonology, morphology, and morphosyntax. Some studies are concerned with the meaning of onomatopoeia, but mostly from semantic or semiotic perspectives. To the best of my knowledge, there has been limited research into onomatopoeia in pragmatics, except for our own observations presented in Sasamoto and Jackson (2016) as well as those of onomatopoeia in Wharton's (2009) work on interjections. Against this backdrop, the aim of this study is to identify the role of onomatopoeia in communication, and how humans communicate using onomatopoeia. Taking a relevance-theoretic approach, I will argue that onomatopoeia involves both verbal and non-verbal aspects of communication, providing direct (non-coded) and direct (coded) evidence for communication. Direct evidence provided by onomatopoeia is based on the perceptual resemblance between the sensory experience the speaker wishes to communicate and the phonological and other linguistic form of the expression. Indirect evidence, however, is what

[3] I will return to Flyxe's analysis in Chaps. 4 and 7.

each onomatopoeic expression encodes. As I will argue in Chap. 4, the encoded element of onomatopoeia may be modified through the lexical modification process of broadening and/or narrowing, to suit a specific context to form an ad hoc concept. This claim might appear similar to the analyses carried out by Kita (1997) and Dingemanse and Akita (2016, 2019), since both analyses assume that two representational dimensions are involved in onomatopoeia. However, the differences between my analysis and previous studies lie in the way in which I see onomatopoeia as a result of humans taking advantage of phonological form and its resemblance to the phenomenon they wish to communicate, rather than the non-arbitrary, systematic link, or iconicity that establishes a form and meaning association, as discussed in many previous studies.

2 Structure of the Book

So far, I have discussed how onomatopoeia has been defined in previous studies, as well as the specific focuses in terms of onomatopoeia research. In Chap. 2, I will turn to onomatopoeia and sound symbolism, where I present an overview of the rich body of work that concerns the iconicity of onomatopoeia and sound symbolism. Drawing on findings from previous studies, this chapter attempts to capture the core assumptions in the sound-symbolism approach. Sound symbolism assumes a non-arbitrary relationship between sound and meaning. Therefore, researchers in this area have often taken an experimental approach to identifying characteristics of sound symbolism, whereby examples of onomatopoeia were presented in pairs to individual subjects in order for them to make a binary selection. Most experiments were in a laboratory setting, rather than a real communicative situation, with the subjects being presented only with words and not in context. I will argue that merely describing the non-arbitrariness of a relationship between sound and meaning does not explain how and why it comes about. I will also point out that the previous definition of *onomatopoeia* is too narrow, and that introducing other terms such as *phonomimes* or *psychomimes* is not necessary when we approach the use of onomatopoeia in communication. Such a classification assumes that the hearer will process onomatopoeia differently from ordinary language use, which would then require a separate account for the interpretation of utterances that contain onomatopoeia. This chapter therefore introduces one of the fundamental questions that I am trying to address in this study: How is onomatopoeia linked to multimodal sensory experience? And where is onomatopoeia located in the intersection between verbal and non-verbal communication?

In Chaps. 3 and 4, I will present a relevance-theoretic account of onomatopoeia as an alternative to a sound-symbolism-based account. In Chap. 3, based on Sasamoto and Jackson (2016), I will argue that onomatopoeia is located on the continuum of *showing* and *saying*, thus providing direct evidence to representation. As already mentioned, research on onomatopoeia is dominated by sound-symbolism approaches, where the focus is on the non-arbitrary relationship between sound and meaning. While this kind of analysis provides a rich description of onomatopoeia in broad terms, it does not account for how and why a speaker uses onomatopoeia, and how a hearer recovers the intended meaning. In addition, as most studies focus on established onomatopoeia, they do not provide any explanation for the motivation behind coining novel or creative onomatopoeia, the mechanism behind comprehension, and notably, as observed in Sasamoto and Jackson (2016), the process in which novel expressions become adapted as fully lexicalised items. I will argue that onomatopoeia should be analysed in communicative terms, in order to gain a full understanding of how it contributes to human communication. To this end, I will show how onomatopoeia provides both direct and indirect evidence to the hearer in communication—that is, *showing* the sensory experience as well as *saying*—thus locating itself on the continuum of *showing–saying*, as well as on the continuum of verbal and non-verbal. This communicative approach to onomatopoeia shows how so-called sound symbolism, or the non-arbitrary relation between sound and meaning, can be explained as a result of the communicator's attempt to recreate their sensory experience by virtue of resemblance.

Having discussed how onomatopoeia involves both direct (non-verbal) and indirect (coded) evidence for communication, I will focus on the linguistic (or coded) aspect of onomatopoeia in Chap. 4, where I will explore possible links between communicative accounts of onomatopoeia and the rich relevance-theoretic literature on lexical pragmatics. I will discuss the degree of lexicalisation of onomatopoeia (cf. Flyxe 2002 and Kadooka 2005) and will argue that what onomatopoeia communicates is, in many cases, non propositional, and that it is not always possible to pin down exactly what it encodes. I will then present an analysis of the meaning of onomatopoeia in the context of lexical pragmatics. I will show how the relevance-theoretic approach to lexical pragmatics can provide an explanation for how the intended meaning of onomatopoeia is recovered. I will argue that what appears to be the polysemous nature of onomatopoeia can be explained in terms of the formation of ad hoc concepts that result from lexical adjustment, such as the narrowing and/or broadening of the encoded concept to suit a particular context. For example, the concept communicated by the use of the onomatopoeic expression *sizzle* includes not just "the hissing sound", but also some kind of sensory impression. This additional layer of information can be explained in terms of the ad hoc concept SIZZLE*, constructed by adjusting the encoded concept of *sizzle* in a particular context. However, the novel and creative onomatopoeia may not have a clear enough concept upon which ad hoc concepts can be constructed, which suggests that onomatopoeia exists in an interesting intersection between the verbal and non-verbal modes, exhibiting the continuum from nonce words to established lexical items.

Having established a relevance-theoretic account of onomatopoeia, I will turn to the more culturally bound nature of onomatopoeia. In Chap. 5, I will focus on the role of onomatopoeia in sensual and creative discourse. In particular, following on from insights into synaesthesia, I will examine a range of onomatopoeia uses in food discourse.

Synaesthesia, or "crossing-over" of the senses, has recently started to attract the attention of scholars working on onomatopoeia (Akita 2013). Of course, synaesthesia is not a communicative phenomenon, whereas onomatopoeia is—synaesthesia involves idiosyncratic (and involuntary) links between cognitive domains, while onomatopoeia can be voluntarily exploited by speakers. However, the common characteristic between onomatopoeia and synaesthesia is that both could involve sensory and linguistic processing, suggesting that onomatopoeia might result from the speaker exploiting connections between sensory domains and linguistic processing for communicative effect, in a way that is similar to how individuals with synaesthesia create such links. Continuing from previous chapters, I will discuss how onomatopoeia in written discourse is a type of enforced synaesthesia. The act of reading is itself an enforced synaesthesia, where what is aural is made visual. Furthermore, in certain genres such as food writing, attempts are made to recreate sensory experience in the mind of the reader. I will argue that onomatopoeia, as the link between sensory experience and linguistic processing, enables enforced synaesthesia and thus facilitates creative links between cognition, affect, and perception.

Following the discussion of onomatopoeia in a creative discourse, in Chap. 6 I will turn to the use of onomatopoeia in the Japanese media. I will discuss how Japanese culture, which is heavily visually oriented, shows rich uses of onomatopoeia. To gain an insight into the interaction between sound and non-sound elements, I will examine a range of onomatopoeia uses from Japanese manga (comic books) and will conclude that onomatopoeia, while possibly being culturally bound, ensures that a full range of multimodal channels are utilised for enhanced communication by binding elements in multimodal discourse. That is, I will argue that onomatopoeia is a bridge between verbal and non-verbal elements in multimodal media.

Finally, in Chap. 7, I will turn to the issue of translating onomatopoeia. The fact that onomatopoeia is so intricately linked to our sensory experience indicates that onomatopoeia is culturally bound. And indeed, it is often argued that the translation of onomatopoeia poses particular difficulties for translators. In this chapter, I will take a corpus-based approach and investigate issues in translating onomatopoeia using a parallel corpus built with user-generated Cookpad data. A range of translation strategies used in translating onomatopoeia will be examined and analysed in terms of the consideration of relevance. I will conclude that the corpus data does not necessarily prove that translating onomatopoeia poses a particular challenge for translators. I will argue that a misconception about the difficulties of translating onomatopoeia stems from seemingly common occurrences of omission of onomatopoeia from the target text. I will also show how such omission is the result of consideration for context, in order to minimise unnecessary cognitive effort and hence optimise relevance.

Bibliography

Akita, Kimi. 2013. Onomatope/Otoshocho No Kenkyu-Shi. In *Onomatope Kenkyu No Shatei: Chikazuku Oto to Imi*, ed. Kazuko Shinohara and Ryoko Uno, 333–364. Tokyo: Hitsuji Shobo.

Akita, Kimi, and Mark Dingemanse. 2019. Ideophones (Mimetics, Expressives). In *Oxford Research Encyclopedia of Linguistics*, ed. Mark Aronoff. Oxford: Oxford University Press.

Akita, Kimi, and Natsuko Tsujimura. 2016. Mimetics. In *Handbook of Japanese Lexicon and Word Formation*, ed. Taro Kageyama and Hideki Kishimoto, vol. 3, 133–160. Berlin and Boston: De Gruyter Mouton. https://doi.org/10.1515/9781614512097-008.

Asano, Tsuruko, ed. 1978. *Giongo-Gitaigo Jiten [The Phonomime Phenomime Dictionary]*. Tokyo: Kadokawa Shuppan.

Atoda, Toshiko, and Kazuko Hoshino. 1995. *Giongo/Gitaigo Tukaikata Jiten. [Usage Guide to Japanese Onomatopoeias]*. Tokyo: Sotakusha.

Childs, G.Tucker. 1995. African Ideophones. In *Sound Symbolism*, ed. Leanne Hinton, Johanna Nichols, and John Ohala, 178–204. Cambridge and New York: Columbia University Press. https://doi.org/10.1017/CBO9780511751806.013.

Diffloth, Gérard. 1972. Notes on Expressive Meaning. *Chicago Linguistic Society* 8: 440–447.

Dingemanse, Mark. 2011a. Ezra Pound among the Mawu: Ideophones and Iconicity in Siwu. In *Semblance and Signification*, ed. Pascal Michelucci, Olga Fischer, and Christina Ljungberg, 39–54. Amsterdam: John Benjamins Publishing. https://www.mpi.nl/publications/446081.

———. 2011b. *The Meaning and Use of Ideophones in Siwu—Max Planck Institute for Psycholinguistics*. Nijmegen: Radboud University. https://www.mpi.nl/publications/1005623.

———. 2011c. Ideophones and the Aesthetics of Everyday Language in a West-African Society. *The Senses and Society* 6 (1): 77–85. https://doi.org/10.2752/174589311X12893982233830.

———. 2012. Advances in the Cross-linguistic Study of Ideophones. *Language and Linguistics Compass* 6 (10): 654–672. https://doi.org/10.1002/lnc3.361.

Dingemanse, Mark, and Kimi Akita. 2016. An Inverse Relation Between Expressiveness and Grammatical Integration: On the Morphosyntactic Typology of Ideophones, with Special Reference to Japanese. *Journal of Linguistics* 53 (3): 501–532. https://doi.org/10.1017/S002222671600030X.

Doke, Clement M. 1935. *Bantu Linguistic Terminology*. London: Longmans. https://www.cambridge.org/core/journals/africa/article/bantu-linguistic-terminology-by-c-m-doke-london-longmans-green-co1935-pp-v-237-6s-net/62F6964DDFCA3941CE2E2E241536FFF5.

Flyxe, Martin. 2002. Translation of Japanese Onomatopoeia Into Swedish (with Focus on Lexicalization). *Africa & Asia* 2: 54–73.

Hamano, Shoko. 1998. *The Sound-Symbolic System of Japanese*. Stanford, CA: CSLI Publications. http://ci.nii.ac.jp/ncid/BA38075516.

Haryu, Etsuko, and Lihua Zhao. 2007. Yuuseion-to Museion-o Daisyoo-Ni Taioo-Zukeru Kankaku-No Kigen: Giongo-Rikai-No Nit-/Tyuu-Hikaku [Understanding the Symbolic Values of Japanese Onomatopoeia: Comparison of Japanese and Chinese Speakers]. *Japanese Psychological Research* 78: 424–432.

Hinton, Leanne, Johanna Nichols, and John Ohala, eds. 1995. *Sound Symbolism*. Cambridge: Cambridge University Press. https://doi.org/10.1017/CBO9780511751806.001.

Ibarretxe-Antunano, Iraide. 2006. *Sound Symbolism and Motion in Basque*. Munich: LINCOM.

Imai, Mutsumi, Sotaro Kita, Miho Nagumo, and Hiroyuki Okada. 2008. Sound Symbolism Facilitates Early Verb Learning. *Cognition* 109 (1): 54–65. https://doi.org/10.1016/j.cognition.2008.07.015.

Inoue, Kazuko. 2013. Onomatope No Tagisei to Sōzōsei [Polysemy and Creativity of Onomatopoeia]. In *OnomatopeKenkyū No Shatei: ChikazukuOto to Imi [Sound Symbolism and Mimetics: Rethinking the Relationship Between Sound and Meaning in Language]*, ed. Kazuko Shinohara and Ryoko Uno, 203–216. Tokyo: Hitsuji Shobo.

Kadooka, Kenichi. 2005. On the Degree of Lexicalization in English Onomatopoeia from a Historical Perspective. *The Ryukoku Journal of Humanities and Sciences* 27 (1): 1–13.

Kakehi, H., I. Tamori, and L. Schourup. 1996. *Dictionary of Iconic Expressions in Japanese*. Berlin: Mouton de Gruyte.

Kantartzis, Katerina, Mutsumi Imai, and Sotaro Kita. 2011. Japanese Sound-Symbolism Facilitates Word Learning in English-Speaking Children. *Cognitive Science* 35 (3): 575–586.

Kita, Sotaro. 1997. Two-Dimensional Semantic Analysis of Japanese Mimetics. *Linguistics* 35 (January): 379–416. https://doi.org/10.1515/ling.1997.35.2.379.

Kwon, Nahyun. 2018. Iconicity Correlated with Vowel Harmony in Korean Ideophones. *Laboratory Phonology: Journal of the Association for Laboratory Phonology* 9 (1): 1. https://doi.org/10.5334/labphon.53.

Laing, Catherine E. 2014. A Phonological Analysis of Onomatopoeia in Early Word Production. *First Language* 34 (5): 387–405. https://doi.org/10.1177/0142723714550110.

———. 2017. A Perceptual Advantage for Onomatopoeia in Early Word Learning: Evidence from Eye-Tracking. *Journal of Experimental Child Psychology* 161: 32–45. https://doi.org/10.1016/j.jecp.2017.03.017.

Laing, Catherine. 2018. A Role for Onomatopoeia in Early Language: Evidence from Phonological Development. *Language and Cognition* 11 (2): 1–15.

Laing, Catherine E. 2019. Phonological Motivation for the Acquisition of Onomatopoeia: An Analysis of Early Words. *Language Learning and Development* 15 (2): 117–197.

Lawson, Nigella. 2001. Creme Brulée. *Nigella.Com*. http://www.nigella.com/recipes/creme-brulee.

Maurer, Daphne, Thanujeni Pathman, and Catherine J. Mondloch. 2006. The Shape of Boubas: Sound–Shape Correspondences in Toddlers and Adults. *Developmental Science* 9 (3): 316–322. https://doi.org/10.1111/j.1467-7687.2006.00495.x.

McCawley, James D. 1968. *The Phonological Component of a Grammar of Japanese*. Hague and Paris: Mouton.

Mithun, Marianne. 1982. The Synchronic and Diachronic Behavior of Plops, Squeaks, Croaks, Sighs, and Moans. *International Journal of American Linguistics* 48 (1): 49–58.

Nuckolls, Janis B. 1996. *Sounds Like Life: Sound-Symbolic Grammar, Performance, and Cognition in Pastaza Quechua*. New York: Oxford University Press.

Nygaard, Lynne C., Allison E. Cook, and Laura L. Namy. 2009. Sound to Meaning Correspondences Facilitate Word Learning. *Cognition* 112 (1): 181–186. https://doi.org/10.1016/j.cognition.2009.04.001.

Ono, Masahiro. 2007. *Nihongo Onomatope Jiten [Japanese Onomatopoeia Dictionary]*. Tokyo: Shogakukan Publishing.

Parault, Susan J., and Paula J. Schwanenflugel. 2006. Sound-Symbolism: A Piece in the Puzzle of Word Learning. *Journal of Psycholinguistic Research* 35 (4): 329–351.

Sasamoto, Ryoko, and Rebecca Jackson. 2016. Onomatopoeia—Showing-Word or Saying-Word? Relevance Theory, Lexis, and the Communication of Impressions. *Lingua* 175: 36–53.

de Saussure, Ferdinand. 1959. *Course in General Linguistics*. New York: Philosophical Library.

Sohn, Ho-Min. 1994. *Korean*. Vol. 25. London and New York: Routledge.

Voeltz, F.K. Erhard, and Christa Kilian-Hatz, eds. 2001. *Ideophones*. Amsterdam: John Benjamins Publishing. https://benjamins.com/catalog/tsl.44.

Wharton, Tim. 2009. *Pragmatics and Non-Verbal Communication*. Cambridge: Cambridge University Press. https://doi.org/10.1017/CBO9780511635649.

Wrembel, Magdalena. 2010. Sound Symbolism in Foreign Language Phonological Acquisition. *Research in Language* 8 (1): 1–14. https://doi.org/10.2478/v10015-010-0013-6.

Yoshida, Hanako. 2004. *Iconicity in Language Learning: The Role of Mimetics in Word Learning Tasks*. PhD Dissertation, Indiana University.

Yoshida, Hanako, and Linda B. Smith. 2003. Sound Symbolism and Early Word Learning in Two Languages. *Proceedings of the Annual Meeting of the Cognitive Science Society* 25 (25): 1287–1292.

Zwicky, Arnold M., and Geoffrey K. Pullum. 1987. Plain Morphology and Expressive Morphology. *Annual Meeting of the Berkeley Linguistics Society* 7 (September): 330–340. https://doi.org/10.3765/bls.v13i0.1817.

2

Onomatopoeia and Sound Symbolism

1 Onomatopoeia and Sound Symbolism

Research on onomatopoeia has been dominated by the sound-symbolism approach.[1] The underlying assumption of this approach is that there is a systematic relationship between sound and meaning. Hinton et al.'s (1995b) edited volume, which is one of the pioneering works in sound symbolism, shows how such "rightness" or "fitness" of sound and meaning correspondence sparked the curiosity of researchers, causing them to focus on identifying the role of sound symbolism in a theory of language. According to Hinton et al. (1995a, 1), "[s]ound symbolism is the direct linkage between sound and meaning". As illustrated in examples (1) to (4) in Chap. 1, there is undeniably some connection between a sound and what that sound is used to describe. That is, little objection would be made to the use of the phonomime (or an onomatopoeic expression that mimics the sound) /*ting*/ to describe the sound of bells. The question is, how do we explain the intuition we have about the link between the sound and the meaning? After all, onomatopoeia does *fit* the entity it is used to describe.

[1] See Akita (2013b) for a fuller discussion of research development in sound-symbolism studies.

R. Sasamoto, *Onomatopoeia and Relevance*, Palgrave Studies in Sound,
https://doi.org/10.1007/978-3-030-26318-8_2

Interest in the relationship between sound and meaning began with the observation that there are sometimes cases where "sound fits images". The idea of sound symbolism is that there is "synesthetic associations between sounds and sensory experience" (Kagitani et al. 2014, 2871), where the link between the phonological form of a word and its meaning appears to be completely natural and non-arbitrary. As such, some claim that the existence of sound symbolism raises questions for the widely accepted views on the arbitrariness of language as proposed by de Saussure (1959). Indeed, there is something that makes such cross-modal mapping feel appropriate. Therefore, the curiosity about this feeling of *fitness* between the sound and meaning, or "the naturalness and rightness of words" (Hinton et al. 1995a, 5), became the starting point for studies that take the sound-symbolism approach (cf. Ramachandran and Hubbard 2001; Kagitani et al. 2014; Akita 2009; Hamano 1998; Hinton et al. 1995a; Ahlner and Zlatev 2010, to name but a few). In the sound-symbolism approach, many researchers assume sound symbolism to be an underlying mechanism of onomatopoeia/mimetics, either explicitly or implicitly.

One of the best-known examples of this approach is Köhler's (1929) famous experiment (later repeated by Ramachandran and Hubbard in 2001) using the made-up words *kiki* and *bouba*. Köhler (1929) and Ramachandran and Hubbard (2001) show how the reactions of participants were universal across languages, when shown two distinct shapes (round and jagged) and two distinct novel onomatopoeia (*kiki* and *bouba*), in that 95% of participants chose the round shape for *bouba* and the jagged shape for *kiki*. This experiment, among others, forms the basis for the claim that the crossover of sensory modes is the underlying mechanism for sound symbolism, in the same way as synaesthesia. That is, sound symbolism is a case of sound (experienced via auditory sense) evoking sensory experience in other senses. Ramachandran and Hubbard (2001) go on to propose that such symbolism plays a pivotal role in the evolution of language, which "provided the initial impetus for language evolution" (Ramachandran and Hubbard 2001, 19). Other scholars (Shinohara and Kawahara 2013; Assaneo et al. 2011, etc.) make similar arguments that sound symbolism is an embodiment of language. This explains how mimetics often involve physical motivation, such as the extent to which the mouth is open, the location of the tongue, or voiced/voiceless tone.

On the assumption that sound symbolism is the "persecutor of fully formed human language" (Hinton et al. 1995a, 11), the focus of sound-symbolism research has been that of the general characteristics and universality of sound-symbolic form and meaning, and, more recently, the cross-modal correspondence of senses (see Chap. 5). Sound-symbolic forms are mostly language specific, hence there is a huge body of literature on the analysis of language-specific sound-symbolic forms (e.g. Akita 2013b; Nasu 2015; Newman 2001; Ibarretxe-Antunano 2006; Ivanova 2006; Nuckolls 2010; and Dingemanse 2011a, b, c). However, as noted by Hinton et al. (1995a, 8–10), onomatopoeia in general tends to involve reduplication (e.g. *ding dong* in English, *betabeta* in Japanese), a marked use of segments and suprasegments (cf. Rhodes 1995), and the association of certain phoneme classes with a particular semantic field. For example, fricatives are often used for quick, audible motions, while continuants are used for continuing sounds and acts, high-frequency consonants are often used for smallness or sharpness, and so on.

While the range of phenomena discussed in sound-symbolism research varies, reviewing them all is beyond the scope of this study. However, the typology of sound symbolism proposed by Hinton et al. (1995a) provides a broad picture of what has been studied in the area. Hintonet al. (1995a) have identified four categories of sound symbolism, which seem to form a continuum of directness in terms of the linkage between sound and meaning.

The first category is *corporeal sound symbolism* or kinaesthesia. According to Hinton et al. (1995a), corporeal sound symbolism represents the internal state of the speaker (emotional or physical). It includes items that are linked to the emotional or physical state of the speaker, such as a cough, a hiccup, an expressive voice quality and intonation, or an interjection. These items are structurally simple and have no written convention *per se*. The classification of Hinton et al. (1995a) includes natural signs as well as the interjection, which seemingly indicates that this class includes both natural meaning (meaningN) and non-natural meaning (meaningNN) in relevance-theoretic terms (see Wharton 2003 and 2009 for a fuller discussion on meaningN and meaningNN).

The second category of sound symbolism proposed by Hinton et al. (1995a) is *imitative sound symbolism*. This category is also called *phonaesthesia*, or *conventional phonetic iconicity*. Imitative sound symbolism is language specific and involves words that mimic environmental sounds (*bang*, *meow*, etc.) external to the speaker. Its similarity to corporeal sound symbolism makes it difficult to establish the spelling for expressions that belong to both corporeal sound symbolism and imitative sound symbolism, but can be conventionalised (Hinton et al. 1995a, 3). The difference between this and corporeal sound symbolism is that it has no link to emotion or physical state. Instead, it has a referential role to the objects which it is used to describe, and often has an internal grammar (cf. Oswalt 1995). This category of sound symbolism also includes movement imitatives. These are phrases such as *ding-dong*, where the reduplication of sounds mimics the rhythmic movement. According to Hinton et al. (1995a, 4), "[j]ust as humans are capable of translating rhythmic sounds into rhythmic movement, they are also capable of the reverse: translating rhythmic movements into sounds, including sound-symbolism language forms". We will return to this particularly interesting point in Sect. 2.

Extending from imitative sound symbolism, Hinton et al. (1995a) present the third category, *synesthetic sound symbolism*. This is also called *synesthetic phonetic iconicity* and, unlike imitative sound symbolism, is considered to be universal. As we will see in Chap. 5, synaesthesia is a condition whereby a stimulus onto one sense triggers a sensory experience in another sensory channel. For example, one might taste something on hearing a particular phoneme. Expressions included in this category seem to exhibit "the process whereby certain vowels, consonants, and suprasegmentals are chosen to consistently represent visual, tactile or proprioceptive properties of objects such as size or shape" (Hinton et al. 1995a, 4), such as palatal consonants and high vowels for diminutives, and a deep voice and lengthening of vowels for large objects (e.g. *biiig fish*).

The last category proposed by Hinton et al. (1995a) is *conventional sound symbolism*, which, they argue, seems to be more established and an integral part of "language", such as an association of certain phonemes and clusters with certain meanings (e.g. *gl* for glitter). Conventional sound symbolism is often language specific and as a result seems closer to arbitrariness, where humans create links with sound and meaning without intrinsic connections in between, as with other sound symbolism. Hinton et al. (1995a) explain that the nature of conventional sound symbolism allows for creating words which are effective in social contexts such as an advertisement. Examples of this kind include brands such as *L'Oréal* for shampoo, where the feminine name *Laura* and the synesthetic sound symbolism for "flowing" sound, symbolising flowing hair, are combined (Hinton et al. 1995a, 6).

Hinton et al. (1995a) present two further categories: *metalinguistic symbolism* and *metacommunicative symbolism*. Metalinguistic symbolism involves items such as segment choice and intonation patterns signalling aspects of linguistic structure and function (e.g. goose vs geese, or vowel ablaut to represent active/passive). Metacommunicative symbolism goes beyond what is traditionally being analysed in terms of sound symbolism. On metacommunicative symbolism, Hinton et al. (1995a) argue that whispering is an example of symbolism, where the acoustic form of speech is adjusted to suit the communicative function. In the case of whispering, the communicative function would be to suggest intimacy or restraint, which is symbolised by the lowered and hushed tone.

The typology proposed by Hinton et al. (1995a) is not the only typology of sound symbolism and there are a number of classifications and subcategories. Japanese, which has a rich repertoire of onomatopoeia, has its own typology of onomatopoeia: *giongo* (phonomimes), *gitaigo* (phenomimes), and *gijogo* (psychomimes); see, for example, Akita and Tsujimura 2016 for a fuller description of these categories. The term *phonomime*—*giseigo* (animate)/*giongo* (inanimate) in Japanese—is used where onomatopoeic expressions mimic the sound, as illustrated in example (1). Onomatopoeic expressions that mimic a manner, state, or condition of an external phenomenon, rather than a sound, are called *phenomimes* (*gitaigo*), as illustrated in (2). In contrast, *gijogo* is particularly interesting, as it is used to describe someone's psychological and emotional state. *Gijogo* is not particularly common in English but is quite prevalent in Japanese. A typical example of *gijogo* is given in (3):

(1) *Giongo* (phonomimes)
Doa ga *gii* tto aita.
Door SUB ONO QUO open-PAST
"The door creaked open."
(2) *Gitaigo* (phenomimes)
teeburu ga abura de *betabeta* shiteiru
table SUB grease with ONO do-ing
"The table is sticky with grease."
(3) *Gijogo* (psychomimes)
Purezento o *wakuwaku* shinagara aketa
Present ACC ONO doing opened
"[I] opened the gift, excited with anticipation."

In (1), the onomatopoeic expression *gii* mimics the sound made by the door, while the phenomime *betabeta* in (2) express the somatosensory quality of the table. *Wakuwaku* used in (3), which is often categorised as *gijogo* or a psychomime, conveys the person's internal emotion/feeling.

While the typology of onomatopoeia might capture the source of symbolism (i.e. what the onomatopoeic expression is used to describe), it is not entirely clear if we need such distinctions, especially when we are concerned with the role of onomatopoeia in communication. Would a speaker process an utterance differently if an onomatopoeic expression contained in the utterance was a phonomime, phenomime, or psychomime? In other words, is the identification of the type of symbolism a prerequisite for the recovery of the intended interpretation of the utterance? Whether an onomatopoeic expression mimics a sound or a feeling, the core of onomatopoeia is its resemblance to the original phenomenon.[2] In addition, as frequently acknowledged by scholars taking a sound-symbolism approach, onomatopoeia often involves various types of semantic extension, including those that are based on a cognitive-linguistic conception of metaphor and metonymy (cf. Tsujimura 2001; Inoue 2013). Furthermore, as we saw earlier, Hinton et al. (1995a) include the case of movement imitative, where the rhythmic movement is translated into sounds. If humans are capable of *translating* a kinaesthetic mode to an auditory mode, it is reasonable to assume that other

[2] I will return to this point in Chap. 3.

cross-modal transfers are also possible. The fact that one onomatopoeic expression can be used to describe an object across different senses suggests that the typology of onomatopoeia, although providing a rich description of what humans are capable of doing in terms of resemblance, does not capture the full picture of its use in real life, let alone explain its role in communication.

The question, then, is how we could explain the role of onomatopoeia in communication. To answer this, we need to address the explanatory gap in existing studies that account for the rich repertoire of symbolism humans can make, as described in the studies that take the sound-symbolism approach. We will come back to this point in the next chapter. But first, we will discuss whether the underlying notion of the systematic link between sound and meaning can be maintained, as generally assumed in sound-symbolism-based studies.

2 Onomatopoeia and Meaning: Systematic Relations?

As discussed in the previous section, it is commonly assumed among scholars taking the sound-symbolism approach that there is a systematic link between sound and meaning. In this section, we will see how far such links can be maintained.

It is generally acknowledged that onomatopoeia is often used in genres that involve highly emotional and sensual language use, such as literature, food discourse, or music critique. In these types of writing, readers are frequently required to go beyond the meaning that is readily available to them. For example, when they read *tinkle* in the poem "The Bells" (see Chap. 1), readers do not just interpret the ringing of the bell. The use of the *onomatopoeia tinkle* also evokes certain sensual impressions that are associated with this particular manner of bell-ringing—not *ding-donging*, not *jingling*, but *tinkling*. It might be the delicate manner in which the bell rings or the faintness of the sound. It would be extremely difficult to spell out exactly what *tinkle* means, but it communicates some really nebulous impression all the same. The question here is how the sound-symbolism approach could account for the intangible nature of what onomatopoeia communicates. Of course, there is no denying that we, as humans, find some sense of fitness between onomatopoeic expressions and what they are used to describe. However, the question is whether such a systematic link between sound and meaning can be maintained, and what ties each sound and meaning.

2.1 Onomatopoeia: Not So Simple Links

Let us take the example of *galumph*, a word invented by Lewis Carrol in his famous work *Through the Looking-Glass*:

(4) He went *galumphing* back.
(Lewis Carrol, *Through the Looking-Glass,* my italics)

Galumph is defined in the *OED* as "to gallop heavily; to bound or move clumsily or noisily". This definition would suggest that *galumph* should be used for something heavy, or at least something that is unagile. However, a quick search on the internet shows that the word is used to describe all types of walking style, including that of cats (a children's picture book by Brenda Landsdown), dinosaurs, frogs (a children's song "Galumph went the little green frog"), and even women in high heels. If there is a systematic link between the sound and the sense, how could the sound *galumph* be systematically linked to so many different ways of walking? One could argue that *galumph* is linked to the manner of walking that is heavy, clumsy, and noisy, *relative to the subject*. However, if the manner of walking described by *galumph* was always relative to the subject, it would require inference from the outset, rather than having a core meaning that is linked to the phonological form of the expression. If one took this line of argument, then one would need to explain how a manner of walking could be "relative to the subject" when the sound-symbolism analysis suggests a systematic link between the manner of walking and the sound *galumph*. After all, in cases like *galumph*, the meaning of the onomatopoeia is linked to various sounds, and, as such, the link has to be adjusted according to the type of subject that is walking. If the meaning can be so easily adjusted, the link between sound and meaning might not be so systematic after all.

The previous example of *galumph* might appear to be the odd one out, as one could argue that it is a highly creative onomatopoeic expression. Or, it might be the case that some examples of onomatopoeia are so general, like *galumph* describing the *noisy* manner of walking, that we cannot find universality between phoneme and meaning. However, as the examples in (5) illustrate, humans seem to be capable of making a link with multiple expressions for one object. Let us take the example of high heels. For instance, *Onomatopoeia Book* (Atkins et al.), which is compiled by literary writers, lists *click-clack* as onomatopoeia for high heels, while an online onomatopoeic dictionary, *Written Sound*,[3] lists *tack tack tack* as onomatopoeia for high heels. Further examples of this are given in (5):

(5) Onomatopoeia for high heels

a. Oh, how I longed for the day when I would be allowed to *clip-clop* around in high heels!

(Teresa Maria, "Day 10: About High Heel-Obsession And How To Find The Perfect Pair", my italics)[4]

b. Nancy ducked between the rolling racks and was slipping the jacket over her T-shirt when she heard the rapid *tack-tack-tack of high-heeled* shoes approaching (Carolyn Keene "Intrigue at the Grand Opera", my italics)[5]

The assumption that there is a systematic link between sound and meaning would suggest that high heels are linked to one sound, not four different sounds, at least in one language. It is true that some of these expressions are similar; in which case, why do we need them all?

Similarly, there is a range of onomatopoeia that have near-identical formation but different sounds for what could be argued to be *similar* materials. In both (11) and (12), onomatopoeic expressions are used to communicate the crunchy, sharp sound of hard materials:

(6) Noise of crunchy bones goes *crackety-crack* for miles around.

(Roald Dahl, *The BFG,* my italics)

[3] http://www.writtensound.com/index.php?term=tack+tack+tack. Accessed 20 October 2017.

[4] https://outlandishblog.com/about-high-heels/. Accessed 12 May 2019.

[5] https://hellogiggles.com/fashion/stopped-high-heels. Accessed 20 October 2017.

(7) *Click clack, clicketty clack, clicketty clicketty clack*!

(Benedict Blathwayt, *Night Flight for the Little Red Train,* my italics)

In (6) and (7), we can see *crackety-crack* and *clicketty clack*. Their formation is almost identical, and in terms of the sound the differences are that (6) is with the /r/ sound, while (7) contains the /l/ sound, and that (7) contains the vowel /ɪ/ rather than the /a/ used in (6). As already mentioned, both *crackety-crack* and *clicketty clack* are used to communicate the crunchy, sharp sound of hard materials, with the onomatopoeia in (6) describing crunching bones and that in (7) describing the sound of the train movement. If there indeed is a link between sound and meaning, it would suggest that the train would have a heavier sound than bones. That is, if there indeed was a systematic relation, motivated by a physical aspect of the articulation of the sound, surely the /krak/ sound, which contains a low back vowel /a/, would be more appropriate for the train, and /klɪk/ would be more appropriate for bones, as the sound-symbolism approach suggests that the vowel /a/ requires a more round and open shape of the mouth, evoking a bigger sound, while the high vowel /ɪ/ would be associated with small items. This is not to say that the author should have used different onomatopoeia. Of course, one could argue that each onomatopoeia was deliberately chosen in order to give rise to the intended effect. However, the point is that the so-called systematic link between sound and meaning does not seem to be driving the choice of onomatopoeia.

Furthermore, it is highly unlikely that readers would even realise that there is a more appropriate onomatopoeia than that used in the text. This, again, raises a question for the sound-symbolism approach. It is not clear whether the systematic link between sound and meaning is the key to explaining the role of onomatopoeia in communication. Of course, the complex nature of onomatopoeia is noted by scholars who take the sound-symbolism approach. For example, Akita and Dingemanse (2019) do indeed acknowledge the complex nature of iconicity and how the formation of ideophones can be bound by the rules of a specific language. They also acknowledge that some connotations of ideophones need to be pragmatically enriched. Yet, their recommendation is an experimental approach to identifying iconicity, which is external to humans, rather than attempting to explain how humans make such iconic connections between form and meaning. However, in order to gain an overview of onomatopoeia, including its nature as a linguistic expression as well as its role in communication, it would be necessary to take both an experimental and a communicative approach to presenting a complementary account.

Another question for the sound-symbolism approach raised in these examples is the non-auditory aspect of onomatopoeia. As illustrated in examples (6) and (7), both onomatopoeic expressions *crackety-crack* and *clicketty clack* have an almost identical formation. This top-heavy, snap-ending rhythm communicates the impression of repetitive, continuous noise. This type of onomatopoeia, where the sound is used to represent a non-sound experience (i.e. cross-modal onomatopoeia), is nothing new. Indeed, it has already been acknowledged in sound-symbolism studies. For example, Hinton et al. (1995a) discuss the translation from movement to sound; Hirata et al. (2011) also report phonetic symbolism and its link to visual lightness; and Ward and Simner (2003) and Ward et al. (2006) report the link between sound and taste. However, in cases such as (6) and (7), it is not a simple matter of transferring from one sense to another. The original sound-based onomatopoeic element is retained, while a rhythmic element to the expressions adds to the "feeling" of the expression. The question is where the link exists. Is it the link between sound and meaning, sound and rhythm, or both? If both, will there be a systematic link to every possible combination of sound and rhythm? What is not clear is the theoretical position of these links.

The onomatopoeic expression in example (8), taken from a well-known Japanese children's story, presents a further challenge, as the expression *Doddodo dodō do, dodoōdo, dodō* represents not only sound or rhythm, but also a fully multimodal experience of being exposed to a strong wind:

(8) [Describing the strong wind that is storming through the village]
Doddodo dodō do, dodoōdo, dodō aoi kurumi mo fukitobase
MIM blue chestnuts also blow-away
Suppai Karin mo fukitobase *Doddodo dodō do, dodoōdo, dodō*
sour quince also blow-away MIM

"Doddodo dodō do, dodoōdo, dodō; Blow away the green chestnuts too;
Blow away the sour quince too; Doddodo dodō do, dodoōdo, dodō…."
(Miyazawa Kenji, *Kaze no Matasaburo* [Matasaburo of the Wind] 1934, 1.
English translation by Strong and Colligan-Taylor 2002, my italics)

When humans experience a strong wind, it is not a matter just of the sound, it also involves how it touches the skin, the force they feel when exposed in the wind, and the rhythm of the wind. In other words, *Doddodo dodō do, dodoōdo, dodō* in (8) represents human experience throughout our senses, not just sound or the combination of closely related modes such as sound and rhythm. This shows how our experience does not single out one particular sense. Humans go through cognitive experiences all the time: some might be strongly sound oriented, and others might not. There will be cases where onomatopoeia does not just involve two modalities—it can be a resemblance between a source experience and multiple modes. So far, it is not clear how sound-symbolism-based approaches could account for this, unless each sound and syllable, or whatever mode is involved, has a systematic link with meaning.

Furthermore, it has been reported that the use of onomatopoeic expressions often accompanies gestures (cf. Hosoma 2017). If the sound-symbolism view of the systematic link between sound and meaning were to be maintained, such a sound-symbolic link would have to be expanded to include accompanying gestures or other modes that humans use to communicate their experience. In the next section, we will see another challenge for the sound-symbolism approach; that is, the role of context in interpreting onomatopoeia.

2.2 Onomatopoeia: The Role of Context

As frequently reported, onomatopoeia is often seen as communicating different meanings, depending on the context in which it is used. *Dondon* in example (9) illustrates this clearly:

(9a) Taro wa *dondon* to doa o tataita.
Taro TOP MIM QUO door ACC hit
"He banged the door."
(9b) Iroirona koto ni *dondon* chosen-shite hoshii.
various matter to MIM challenge-do want
"I would like you to challenge various new things."

In (9a), *dondon* represents the sound Taro is making by banging the door. In contrast, *dondon* in (9b) has little to do with the actual sound, but is used to describe the manner in which a person overcomes different challenges in life, one after the other. Both uses of *dondon* in (9) might share the feeling of continuity (of the banging and of the life journey) and the forceful manner of the action involved. However, what *dondon* describes in (9a) is an immediate noise in a particular situational context, while *dondon* in (9b) describes the less concrete action of a person in a longitudinal context. Some scholars (cf. Tsujimura 2001; Akita 2013a, c; Inoue 2013) attempt to account for this *polysemous* nature of onomatopoeia in terms of *semantic extension*.[6] This is a notion from frame semantics which attempts to explain the links between meanings of an apparently polysemous lexical item, in order to "locate the sound symbolism research in the center of cognitive studies of language" (Akita 2013c, 1). In frame semantics (Fillmore 1982, 1985), words can be categorised into groups based on a semantic frame. Frame elements have core elements that must be shared by all members of the group and non-core elements that are optional. Akita (2013c) analyses onomatopoeic expressions that represent voices and noises from four languages (English, Japanese, Korean, and Mandarin Chinese) and proposes three constraints on the semantic extension of onomatopoeia: namely, referential specificity, event-semantic complexity, and syntactic status. Akita (2013c) argues that these constraints determine the semantic extensibility of onomatopoeic expressions. These constraints can either be language specific or common across languages and are consistent with the role of frame semantics. Akita

[6] In relevance theory, on which this study is based, metaphor or metonymy interpretation is seen as a pragmatic process, not a semantic process. I will return to the notion of semantic extension in Chap. 4.

(2013c) explains that the low semantic extensibility (or referent neutrality) of Chinese and English is a result of the absence of specific frame elements. He also argues that the event-structural constraint accounts for the resistance of onomatopoeic expressions for voice (as opposed to noise), as such expressions involve a simple frame, while onomatopoeic expressions for noises often involve more complex frames, sometimes including a subevent (subframe). Finally, he observes the syntactic constraint, the strongest constraint of the three, where he finds the correlation between the morphosyntactic integration and semantic extensibility of onomatopoeic expressions. He explains how linguistic items that take grammatically core positions, rather than adjuncts, have a stronger extensibility. Adjuncts, such as onomatopoeia used with a quotation particle in Japanese, are not part of the core clause and hence are less strongly paired with predicates, leading to less extensibility.

The use of *extension* suggests that there is a base meaning, while other polysemous meanings are derivative of this base meaning. *Semantic extension*, or *metaphor extension* or *metonymy extension* as often analysed, might capture the intuition about the relationship between the meanings of *dondon* in (9a) and (9b). This would suggest that the core element shared by all members of the group is derived by the systematic link between sound and meaning, and that other polysemous meanings would have an optional non-core element that differentiates the meanings of each. The question is, would humans have inventories of such frames for each use of onomatopoeia, rather than the meaning of such an expression being identified (narrowed down or broadened) in the pragmatic process? That is, the sound-symbolism approach, by allocating frame elements to each expression, will overlook the role the context plays in the hearer's interpretation process. As shown in the subsequent examples, one cannot overlook the importance of contextual assumptions. Wilson (2003, 283) notes:

> [U]nderstanding any utterance, literal, loose or metaphorical, is a matter of seeing its intended relevance, and seeing the intended relevance of an utterance is a matter of following a path of least effort in mutually adjusting explicit content, context and cognitive effects, as specified in the relevance-theoretic comprehension procedure.

That is, whether or not some utterance is considered "figurative", it all involves the same interpretation process and can be explained in the same way: in terms of relevance.

The role that context plays is crucial for the interpretation of onomatopoeia. As generally acknowledged in relevance theory, the assumptions that the hearers have, or have access to, depend on the individual. This suggests that different hearers would interpret utterances differently, according to their own ability. As a result, there are cases where successful communication depends on the hearers' contextual assumptions, and therefore some utterance might fail to convey the speaker's intended meaning. This poses a challenge for approaches that overlook the significant role played by context in utterance interpretation. Let me take an example of onomatopoeic expressions to describe the texture of food. Following is an extract of a conversation on a Japanese TV programme, a popular chat-show where a group of celebrity panellists discuss various (often trivial) topics. This frequently takes the form of a talk battle between two groups of panellists with opposing views. The theme of this particular episode was "Women who cannot cook vs men who love cooking". The male panellists in this episode are all famous for being "foodies", whereas the female panellists are uninterested either in cooking or in food in general.

The topic of discussion in this particular scene is fried rice, which is one of the most popular dishes in Japan. A few months prior to the broadcast of this episode, there was another episode in which a big debate on fried rice took place. One of the panellists (Panellist no. 2) is an actor known for publishing a book on fried rice.[7] As we can see from this extract, the programme starts with a debate among male panellists about the texture of fried rice:

(10) Fried rice example (onomatopoeia is marked in italics)
MC: Katsumata (Panellist 1), (how do you like) fried rice?
Panellist 1: I prefer *betabeta*, like the sort you get at a greasy spoon. I'm not really into *parapara*.

[7] Kanayama Kazuhiko ryu Chahan no gokui & Okazu no moto special. [Kazuhiko-Kanayama style, The secret of fried rice & other dishes (my translation)].

MC: Oh, you prefer *betabeta*
Panellist 1: (the kind of) fried rice dinner ladies would make
MC: Ah, got it
Panellist 3: But fried rice is not *parapara*. Not *parapara* but *parapara shittori*. (Fried rice) would not taste nice unless (it was) *parapara shittori*.
MC: How about you, Mr. Ko (Panellist 3)?
Panellist 3: Current (fried rice) is *parapara* and *shittori*.
MC: Current?
Panellist 3: (It has become) mainstream now
Panellist 2: Once upon a time, *parapara* was good too, but now it is *parapara shittori*.
MC: Can you make it so?
Panellist 2: Yes
[...]
MC: Do you want to say something, Mr Ozawa (Panellist 4)?
Panellist 4: Well, but fried rice (is/should be) "*bote*"
Panellist 1: Yes!
[laughter]
Panellist 5: He described (fried rice) with a sound, "*bote*"!
Panellist 4: You see, you do this (gesture of using a wok) and do *bon* [inaudible]
MC: Can you explain, so that we can understand (what) "*bote*" [is]? You mean, something heavy?
Panellist 4: Heavy indeed.
Panellist 3: Fried rice should become *fuwa* when you do *pon*
Panellist 5: No, it is not total nonsense (I can kind of understand) the sound of *bon* when you (flip) the ladle to (serve) fried rice (onto a plate)
MC: How about you, Christine? (Panellist 6)
Panellist 6: I cannot cook at all. I have no idea what you are all talking about, saying things like *parapara* or *betabeta*.
MC: Oh, you do not understand rice being *parapara*, *betabeta* or *don!*
(Extract from "Odoru Sanma Goten", Nihon TV, broadcast 21 June 2016)

As we can see from these examples, male panellists use a number of onomatopoeic expressions to describe their favourite texture of fried rice, including *betabeta*, *parapara*, *shittori*, *parapara-shittori*, *bote*, *don*, *fuwa*, and *pon*. Interestingly, in this example we see the sharp contrast between the male panellist group and the female panellist group. While all the male panellists eagerly engage in the discussion, the female panellists seem uninterested and do not engage in the conversation. As mentioned earlier, the male panellists have a strong interest in food and cooking, while the female panellists have little interest in that topic. It is particularly interesting that this contrast in interest seems to result in a difference in understanding. The utterance by Panellist 6 illustrates this—her lack of interest in food and cooking results in her not being able to follow the discussion. If someone is not particularly interested in a topic, it is not surprising that they would have fewer assumptions in relation to that topic. That is, female panellists in this scene appear to have far more limited contextual assumptions than male panellists. The lack of assumptions about fried rice, whether about the texture, taste, or recipe, causes difficulties in interpreting utterances that would require a considerable amount of knowledge and experience with fried rice, especially when the utterances in question involve a number of onomatopoeia with subtle differences.

This example raises a particular challenge for the sound-symbolism approach. The notion of sound symbolism suggests that one could recover the link between meaning and sound, as such a link is systematic. As acknowledged in the literature, the universality of sound symbolism has been little explored. However, there is a huge body of literature demonstrating language-specific sound symbolism. If it is right to assume that sound symbolism is at least language specific, one would expect proficient speakers of a certain language to be able to recover such a link. However, as the extract above illustrates, even among native speakers of Japanese there are differences in the ability to interpret onomatopoeic expressions.

Another issue that has rarely been discussed in relation to sound symbolism is the interchangeability of onomatopoeia that seem closely related to each other. Earlier, we saw how scholars conducted lab-based experiments where participants were asked to make a binary selection. In many cases, such experiments were done using two contrasting expressions, or in a controlled environment. In real life, however, there are many overlaps where similar onomatopoeia can be used. For example, Sugimura (2017) conducted a survey of 105 participants (81 native speakers of Japanese, 19 advanced learners of Japanese in Japan, and 15 beginners of Japanese in China). The survey focused on the areas of the body, types of injury, and onomatopoeia for pains and bodily feeling, such as *piripiri*, *hirihiri*, *biribiri*. He also conducted a corpus-based study on onomatopoeia and its collocation. The results of his survey are particularly interesting, as it is shown that even native speakers of Japanese failed to make clear distinctions between expressions that appear to have similar meanings. However, Sugimura's corpus analysis shows that there are subtle differences among these expressions. The fact that differences between expressions, however subtle, can be found in real-life examples clearly indicates that meaning in onomatopoeia cannot be explained independently of the context in which each expression is used. This raises further challenges for existing studies, which are often based on lab-based experiments.

3 Conclusion

As we have seen, a huge volume of previous studies takes sound-symbolism-based approaches that claim to reveal the non-arbitrary nature of onomatopoeia and mimetics. It is true that cases of sound symbolism suggest humans are capable of forming a non-arbitrary relationship between sound and meaning. However, as discussed in this chapter, it is not entirely clear how and why they take advantage of such links in communication. More analysis on the use of onomatopoeia in context, rather than in a lab setting, would provide further insight into this. The cases in which hearers cannot recognise meaning from onomatopoeia alone, even in sentences (i.e. the fried rice example in (10)), are evidence that hearers need shared assumptions or multimodal input to help with the recovery of meaning. This suggests that the so-called non-arbitrary relationships between sound and meaning are not enough to claim that sound symbolism forms the basis of language, as some scholars seem to imply. The focus on sound symbolism for a theory of *language* overlooks the fact that it might not be linguistics that determines the *core* of sound symbolism.

Sound symbolism is a result of humans making a link between sound and meaning, and the realisation of such sound symbolism might be bound by certain constraints in specific languages. So, descriptions such as a round mouth being equal to a round object would not yield perfectly universal results. Otherwise, why would we have so many different words across languages for barking? That is, sound symbolism is a result of humans' capacity to express their sensory experience by virtue of the resemblance between the phonological form and the experience. Rather, the question should be how humans make such links—how is onomatopoeia linked to humans' sensory experience? The next chapter will explore this question.

Bibliography

Ahlner, Felix, and Jordan Zlatev. 2010. Cross-Modal Iconicity: A Cognitive Semiotic Approach to Sound Symbolism. *Sign Systems Studies* 38 (1/4): 298–348.

Akita, Kimi. 2009. *A Grammar of Sound-Symbolic Words in Japanese: Theoretical Approaches to Iconic and Lexical Properties of Mimetics*. Kobe: Kobe University.

———. 2013a. Kyoki Tokusei Kara Miru Onomatope No Fureemu Riron. In *Onomatope Kenkyu No Shatei: Chikazuku Oto to Imi* , ed. Kazuko Shinohara and Ryoko Uno, 101–116. Tokyo: Hitsuji Shobo.

———. 2013b. Onomatope/Otoshocho No Kenkyu-Shi. In *Onomatope Kenkyu No Shatei: Chikazuku Oto to Imi* , ed. Kazuko Shinohara and Ryoko Uno, 333–364. Tokyo: Hitsuji Shobo.

———. 2013c. Constraints on the Semantic Extension of Onomatopoeia. *Public Journal of Semiotics* 5 (1): 21–37.

Akita, Kimi, and Mark Dingemanse. 2019. Ideophones (Mimetics, Expressives). In *Oxford Research Encyclopedia of Linguistics*, ed. Mark Aronoff. Oxford: Oxford University Press.

Akita, Kimi, and Natsuko Tsujimura. 2016. Mimetics. In *Handbook of Japanese Lexicon and Word Formation*, ed. Taro Kageyama and Hideki Kishimoto, vol. 3, 133–160. Berlin and Boston: De Gruyter Mouton. https://doi.org/10.1515/9781614512097-008.

Assaneo, Maria Florencia, Juan Ignacio Nichols, and Marcos Alberto Trevisan. 2011. The Anatomy of Onomatopoeia. *PLoS ONE* 6(12): e28317. https://journals.plos.org/plosone/article?id=10.1371/journal.pone.0028317.

Dingemanse, Mark. 2011a. Ezra Pound among the Mawu: Ideophones and Iconicity in Siwu. In *Semblance and Signification*, ed. Pascal Michelucci, Olga Fischer, and Christina Ljungberg, 39–54. Amsterdam: John Benjamins Publishing. https://www.mpi.nl/publications/446081.

———. 2011b. *The Meaning and Use of Ideophones in Siwu—Max Planck Institute for Psycholinguistics*. Nijmegen: Radboud University. https://www.mpi.nl/publications/1005623.

———. 2011c. Ideophones and the Aesthetics of Everyday Language in a West-African Society. *The Senses and Society* 6 (1): 77–85. https://doi.org/10.2752/174589311X12893982233830.

Fillmore, Charles. 1982. Frame Semantics. In *Linguistics in the Morning Calm*, ed. The Linguistics Society of Korea, 111–137. Seoul: Hanshin. https://www.coursehero.com/file/29281284/Fillmore-1982-Frame-Semanticspdf/.

———. 1985. Frames and the Semantics of Understanding. *Quaderni Di Semantica* 6: 222–254.

Hamano, Shoko. 1998. *The Sound-Symbolic System of Japanese*. Stanford, CA: CSLI Publications. http://ci.nii.ac.jp/ncid/BA38075516.

Hinton, Leanne, Johanna Nichols, and John Ohala, eds. 1995a. *Sound Symbolism*. Cambridge: Cambridge University Press. https://doi.org/10.1017/CBO9780511751806.001.

———. 1995b. Introduction: Sound-Symbolic Processes. In *Sound Symbolism*, ed. Leanne Hinton, Johanna Nichols, and John Ohala, 1–12. Cambridge: Cambridge University Press. https://doi.org/10.1017/CBO9780511751806.001.

Hirata, Sachiko, Jun Ukita, and Shinichi Kita. 2011. Implicit Phonetic Symbolism in Voicing of Consonants and Visual Lightness Using Gerner's Speeded Classification Task. *Perceptual and Motor Skills* 113 (December): 929–940. https://doi.org/10.2466/15.21.28.PMS.113.6.929-940.

Hosoma, Hiromichi. 2017. *Japanese Language Education in Europe*. Belfast.

Ibarretxe-Antunano, Iraide. 2006. *Sound Symbolism and Motion in Basque*. Munich: LINCOM.

Inoue, Kazuko. 2013. Onomatope No Tagisei to Sōzōsei [Polysemy and Creativity of Onomatopoeia]. In *OnomatopeKenkyū No Shatei: ChikazukuOto to Imi [Sound Symbolism and Mimetics: Rethinking the Relationship Between Sound and Meaning in Language]*, ed. Kazuko Shinohara and Ryoko Uno, 203–216. Tokyo: Hitsuji Shobo.

Ivanova, Gergana. 2006. Sound-Symbolic Approach to Japanese Mimetic Words. *Toronto Working Papers in Linguistics* 26: 103–114.

Kagitani, Tatsuki, Mao Goto, Junji Watanbe, and Maki Sakamoto. 2014. Sound Symbolic Relationship between Onomatopoeia and Emotional Evaluations in Taste. *Proceedings of the Annual Meeting of the Cognitive Science Society* 36: 2871–2876.

Kenji, Miyazawa, Sarah M. Strong, and Karen Colligan-Taylor. 2002. *Masterworks of Miyazawa Kenji – Poems and Fairy Tales*. Tokyo: International Foundation for the Promotion of Languages and Culture.

Köhler, Wolfgang. 1929. *Gestalt Psychology*. New York: H. Liveright.

Nasu, Akiko. 2015. The Phonological Lexicon and Mimetic Phonology. In *Handbook of Japanese Phonetics and Phonology*, ed. Haruo Kubozono, 253–288. Berlin: Mouton de Gruyter.

Newman, Paul. 2001. Are Ideophones Really as Weird and Extra-Systematic as Linguists Make Them Out to Be? In *Ideophones*, ed. F.K.E. Voeltz and C. Kilian-Hatz, 251–258. Amsterdam: John Benjamins.

Nuckolls, Janis B. 2010. The Sound-Symbolic Expression of Animacy in Amazonian Ecuador. *Diversity* 2 (3): 353–369. https://doi.org/10.3390/d2030353.

Oswalt, Robert L. 1995. Inanimate Imitatives in English. In *Sound Symbolism*, ed. Leanne Hinton, Johanna Nichols, and John Ohala, 293–306. Cambridge: Cambridge University Press. https://doi.org/10.1017/CBO9780511751806.020.

Ramachandran, V.S., and E.M. Hubbard. 2001. Synaesthesia—A Window Into Perception, Thought and Language. *Journal of Consciousness Studies* 8 (12): 3–34.

Rhodes, Richard. 1995. Aural Images. In *Sound Symbolism*, ed. Leanne Hinton, Johanna Nichols, and John Ohala, 276–292. Cambridge: Cambridge University Press. https://doi.org/10.1017/CBO9780511751806.019.

de Saussure, Ferdinand. 1959. *Course in General Linguistics*. New York: Philosophical Library.

Shinohara, Kazuko, and Shigeto Kawahara. 2013. Otoshocho No Gengo Fuhensei: "ookisa" No Imeeji o Motoni [Linguistic Universality of Sound Symbolism: Based on the Images of Size]. In *OnomatopeKenkyū No Shatei: ChikazukuOto to Imi [Sound Symbolism and Mimetics: Rethinking the Relationship Between Sound and Meaning in Language]*, 43–57. Tokyo: Hitsuji Shobo.

Sugimura, Yasushi. 2017. Nihongo No Onomatope "Hirihiri HiriQ, Hiriri", "Piripiri, PiriQ, Piriri", "Biribiri, BiriQ, Biriri" No Kijutsuteki Kenkyu [A Descriptive Study of Japanese Onomatopoeia "Hirihiri HiriQ, Hiriri",

"Piripiri, PiriQ, Piriri", "Biribiri, BiriQ, Biriri"]. *Nagoya Daigaku Gengo Bunka Kenkyukai, Kotoba No Kagaku [Studia Linguistica]* 31: 111–130.
Tsujimura, Natsuko. 2001. Revisiting the Two-Dimensional Approach to Mimetics: Reply to Kita (1997). *Linguistics* 39 (2): 409–418.
Ward, Jamie, and Julia Simner. 2003. Lexical-Gustatory Synaesthesia: Linguistic and Conceptual Factors. *Cognition* 89 (3): 237–261.
Ward, Jamie, Brett Huckstep, and Elias Tsakanikos. 2006. Sound-Colour Synaesthesia: To What Extent Does It Use Cross-Modal Mechanisms Common to Us All? *Cortex* 42 (2): 264–280. https://doi.org/10.1016/S0010-9452(08)70352-6.
Wharton, Tim. 2003. Natural Pragmatics and Natural Codes. *Mind & Language* 18 (5): 447–477. https://doi.org/10.1111/1468-0017.00237.
———. 2009. *Pragmatics and Non-Verbal Communication*. Cambridge: Cambridge University Press. https://doi.org/10.1017/CBO9780511635649.
Wilson, Deirdre. 2003. Relevance and Lexical Pragmatics. *Italian Journal of Linguistics* 15 (January): 273–291.

3

Onomatopoeia, the *Showing–Saying Continuum*, and Perceptual Resemblance

In Chap. 2, I showed how broad approaches to the treatment of onomatopoeia yield a number of insights and empirically supported claims which advance our understanding of onomatopoeia. However, they also raise a number of issues. The biggest concern is that little attention seems to have been paid to the role of onomatopoeia in communication, as most studies concentrate on the classification of onomatopoeia. Studies on onomatopoeia frequently use examples of established (or fully lexicalised) onomatopoeia and their binary selection, in that they are often presented in pairs to individuals, who then make a choice of suitability in context, where expressions with contrasting meaning are paired. The best-known example of such an approach, as mentioned in Chap. 2, would be the use of *bouba* and *kiki* in the famous experiment conducted by Köhler (1929), and repeated by Ramachandran and Hubbard (2001). To my knowledge, little research has been done to investigate how and why humans use onomatopoeia. If research into onomatopoeia fails to address its role in communication, and how the hearer interprets it, our understanding of the phenomenon will not be sufficiently wide-reaching or explanatory.

R. Sasamoto, *Onomatopoeia and Relevance*, Palgrave Studies in Sound,
https://doi.org/10.1007/978-3-030-26318-8_3

Also, as pointed out in Sasamoto and Jackson (2016), another issue raised by the approaches taken in previous studies is the way in which the meaning of onomatopoeia has been treated. Either explicitly or implicitly, it has been suggested that there is a systematic link between particular phonemes (or combinations thereof) and meanings. That is, most studies are engaged with the contribution of onomatopoeia to *semantics*, understood as the study of linguistically encoded meanings. By contrast, there is virtually no study that is concerned with the role of pragmatic inferencing in the interpretation of onomatopoeia, despite the acknowledgement of stylistic effects associated with onomatopoeia (cf. Kita 1997; Toratani 2005, 2013; Tsujimura 2001). While these scholars, either explicitly or implicitly, describe the stylistic effects of onomatopoeia, the studies fail to provide a cognitively grounded explanation for how these effects arise; nor do they explain how and why these effects are recovered.

Therefore, the aim of this chapter is to address this explanatory gap and to account for the interpretation of onomatopoeia in terms of what speakers intend to communicate by it. To this end, this chapter presents a relevance-theoretic analysis of onomatopoeia. This was first presented in Sasamoto and Jackson (2016). In order to account for the role of onomatopoeia in communication and how such "stylistic effects" are recovered, I will first present our analysis from Sasamoto and Jackson (2016) using the relevance-theoretic framework, which is well equipped to allow us to discuss onomatopoeia in cognitive, communicative, and inferential terms (Sperber and Wilson 1986/1995; Carston 2002; Wilson and Sperber 2012).

1 Relevance Theory

1.1 Overview

Relevance theory was first proposed by Sperber and Wilson in 1986. From the beginning, it was ground-breaking in many ways, not least because it acknowledges explicitly that communication is not just a matter of decoding and encoding, or inference alone. Sperber and Wilson (1986/1995) show how communication involves both coding *and* inference. Of course, they were not the first to acknowledge that communication involves more than coding. Grice (1957) famously set the foundations for a theory of inferential communication by making a clear distinction between what is said and what is meant, and he proposed that communication can be explained in terms of the cooperative principle and four maxims: quality, quantity, manner, and relevance. Grice's theory of communication has attracted much criticism, as well as many advocates. One criticism is, of course, the question of, or justification for, the four maxims: what are they and where did they come from?

As the name would predict, it is sometimes claimed that Sperber and Wilson (1986/1995) developed their theory from the maxim of relevance. However, unlike Grice's notion of relevance, Sperber and Wilson (1986/1995) ground their theory firmly in human cognition. Rather than making a list of what might influence communication, Sperber and Wilson (1986/1995) propose that we can explain communication through the lens of human cognition. That is, relevance theory is a cognitively grounded theory of communication.

Relevance theory is centred around two principles: the *cognitive principle of relevance* and the *communicative principle of relevance*. The first, the cognitive principle of relevance, describes how human cognition is developed:

(1) Human cognition tends to be geared to the maximisation of relevance.

(Sperber and Wilson 1986/1995, 260)

As this principle states, our cognition has evolved in such a way that we pay attention to something that might give us some reward. That is, we pay attention to what seems likely to be worth our processing effort and is likely to create cognitive effects. In other words, we pay attention to what is relevant. Cognitive effects are derived when the input we receive changes our representation of the world. Let us consider this scenario. One day, you find a dog that is lost on campus. You agree to take the dog home for a couple of days while looking for the owner. After a few days, you locate the owner, but they refuse to take the dog back, and you decide to officially adopt the dog. From that point, you increasingly start to notice posters on lampposts put up by people looking for their beloved dog that has been lost or stolen, social media campaigns for animal charities, or, simply, sad-looking dogs wandering the streets. This is not because animal charities started the campaign after you adopted the dog. The number of sad-looking dogs did not increase overnight. Moreover, there might not even be new posters on the lampposts. You started to notice these things because dogs that are in need of a new home became relevant to you. They led to cognitive effects in a way that they previously did not. Looking at animal charities' campaign adverts might strengthen your assumption that providing a new home for the dog you found on campus was the right thing for you to have done. Or, it might contradict your existing assumption that all dog owners are responsible, and that the former owner of your dog was a very minority exception. In such a case, your existing assumption would be eliminated from your cognitive environment. Furthermore, when you see a sad-looking dog on the street, you might come to a new conclusion that the government should introduce a licensing system for dog owners, whereby they need to pass a test to be allowed to have a pet dog. All of these are changes in your cognitive environment; that is, cognitive effects derived from what you paid attention to—things that concern dogs in distress are now relevant to you.

Relevance is defined in terms of the balance between cognitive effects and processing effort. All other things being equal, the less effort required to process an input, the more relevant it is. Similarly, all other things being equal, the more cognitive effects derived from processing a stimulus, the more relevant it is. That is, relevance is a balance between processing effort on the one hand and cognitive effects on the other.

This cognitive tendency of humans explains the mechanism of human communication—how and why communication works:

(2) Every act of ostensive communication communicates a presumption of its own optimal relevance.

(Sperber and Wilson 1986/1995, 260)

Sperber and Wilson (1986/1995, 260–278) define optimal relevance as follows:

(3) Optimal relevance

An ostensive stimulus is optimally relevant to an audience iff:

a. It is relevant enough to be worth the audience's processing effort;
b. It is the most relevant one compatible with communicator's abilities and preferences.

Our cognitive tendency, which is stipulated in the cognitive principle of relevance, would suggest that if someone tries to attract your attention, that person would have a good reason to do so. That is, when someone demands your cognitive effort (= attention) by producing an utterance, or any other ostensive communicative stimulus, then the hearer would be automatically entitled to assume that whatever the speaker is trying to communicate must be worth their processing. Otherwise, why would anyone pay attention to anything that is communicated to them? In technical terms, when the hearer recognises the speaker's communicative act, such as producing an utterance, they presume that the utterance is optimally relevant, and therefore they start searching for an interpretation that satisfies this presumption. After all, according to the *communicative principle of relevance* presented in (2), a hearer is entitled to expect a speaker's ostensive act to be at least relevant enough to be worth processing, and, moreover, to be the most relevant one that the speaker is willing and able to produce at that time. This is called the *presumption of optimal relevance*:

(4) Presumption of optimal relevance
a. The ostensive stimulus is relevant enough for it to be worth the addressee's effort to process it.
b. The ostensive stimulus is the most relevant one compatible with the communicator's abilities and preferences.

(Sperber and Wilson 1986/1995, 270)

Let us put this a different way. As Sperber and Wilson (1986/1995, 51–54) explain, there are two layers of information in communication that the hearer must retrieve. The first layer is the information that the speaker intends to communicate, and the second is the information that the first layer of information is being pointed out intentionally. That is, ostensive-inferential communication involves making manifest to the hearer that you intend to make the first layer of information manifest. As a result, upon recognising two layers of information, motivated by the presumption of optimal relevance, the hearer would process the ostensive communicative stimuli such as an utterance. According to relevance theory, the interpretation process is that the hearer follows a comprehension path to derive optimal cognitive effects with the least effort, on recognising the speaker's offering of an ostensive stimulus:

(5) The relevance-theoretic comprehension procedure:
  a. Follow the path of least effort in deriving cognitive effects: test interpretive hypotheses (reference assignments, disambiguations, implicatures, etc.) in order of accessibility.
  b. Stop when your expectations of relevance are satisfied.

(Sperber and Wilson 2002, 24)

As illustrated in (5), we can see that interpreting an utterance involves testing interpretive hypotheses. Testing interpretive hypotheses involves a number of inferential processes, and we do not necessarily test all of them at the same time, some of which might, or might not, be necessary for interpreting a particular utterance.

1.2 Communication by Virtue of Resemblance

One of the revolutionary departures from traditional (Gricean) pragmatics is relevance theory's acknowledgement of the human capacity to (meta)represent thought. Instead of considering that all utterances have a propositional content that can be assigned a truth value, relevance theory acknowledges that not all utterances are used to describe a state of affairs. Sperber and Wilson (1986/1995, 228) show how resemblance, rather than a description of the world, can be exploited in communication:

> Any representation with a propositional form, and in particular any utterance, can be used to represent things in two ways. It can represent some

> state of affairs in virtue of its propositional form being true of that state of affairs; in this case we will say that the representation is a description, or that it is used descriptively. Or it can represent some other representation which also has a propositional form—a thought for instance—in virtue of a resemblance between the two propositional forms; in this case we will say that the first representation is an interpretation of the second one, or that it is used interpretively.

To illustrate this, they examine a range cases of resemblance in (6). In (6), Peter is looking for his wallet which he thinks he has left at the inn. Mary's response to Peter's question in (6a) is a direct quotation of what the inn-keeper said, and she uses this utterance not because it describes the world, but because it resembles what the inn-keeper says. It happens to be an identical match to the linguistic form originally used by the inn-keeper. In (6b), Mary's utterance is a translation of the original utterance. As such, it does not correspond to the linguistic form of the original utterance, but it does correspond to the semantic structure of the original. In contrast, Mary's answer in (6c) does not correspond to the original utterance, in either semantic or linguistic form. Instead, it resembles the original utterance in terms of its propositional form. Finally, in (6d), Mary's utterance does not resemble the original in terms of linguistic, semantic, or propositional form, but resembles its logical form.

(6a6) Resemblance in terms of linguistic structure
Peter: And what did the inn-keeper say?
Mary: Je l'ai cherché partout!

(6b) Resemblance in terms of semantic form
Peter: And what did the inn-keeper say?
Mary: I looked for it everywhere.

(6c) Resemblance in terms of propositional properties
Peter: And what did the inn-keeper say?
Mary: He has looked for your wallet everywhere. I don't believe him, though.

(6d) Resemblance in terms of linguistic structure
Peter: And what did the inn-keeper say?
Mary: That he has looked for your wallet everywhere.

(Sperber and Wilson 1986/1995, 227–228)

Although these examples are all based on the resemblance between representations with a propositional form developed from an utterance, non-verbal stimulus is not necessarily excluded. Imagine this scenario—at a family gathering, you are supervising children away from other grown-ups. After a while, the host notices that you are without a drink, raises his hand to the height of his face and moves his hand closer to his mouth, as if drinking something from the glass he was holding. You would most likely interpret this series of gestures as an offer of a drink. In this scenario, the host uses his gesture that resembles the act of drinking to communicate his intention to offer the drink.

Similarly, communication by virtue of resemblance does not necessarily exclude non-propositional resemblance. In fact, even when an utterance is used by virtue of resemblance to other representations, it is not necessarily concerned with its proposition:

(7) Peter: What language did you speak to the inn-keeper?
Mary: Bonjour, comment allez-vous, bien, merci, et vous?
(Sperber and Wilson 1986/1995, 227)

As Sperber and Wilson (1986/1995) explain, while Mary's utterance in (7) clearly has a propositional form, what she intended to communicate is not its proposition. Instead, what she intended to make manifest is the fact that she spoke to the inn-keeper in French.

These examples illustrate how any phenomenon can be used in communication by virtue of its resemblance to some other phenomenon. That is, any type of resemblance, be it visual, linguistic, semantic, phonetic, or logical, can be used in communication. As long as two objects or actions share certain properties, they resemble each other. On recognising a communicative stimulus used by virtue of resemblance, the hearer will follow the relevance-theoretic comprehension procedure and identify the properties that are shared between the two. The hearer will stop when they reach the interpretation that satisfies the expectation of relevance (i.e. the one that yields enough implications).

The fact that humans have a capacity to entertain representations based on resemblance is enough to suggest that, in addition to interpretive resemblance, they can entertain representations by virtue of non-propositional (and often perceptual) representations. In this sense, the imitative (or iconic) nature of onomatopoeia, whether it be an imitation of sound or a psychological state, would also fall within the category of resemblance. As mentioned in Chap. 1, onomatopoeia ranges from extremely clear to less clear cases in terms of resemblance. It is evident that onomatopoeia based on an extremely clear resemblance in terms of sound, such as *bow-wow*, is an instance of phonological resemblance. Unfortunately, Sperber and Wilson (1986/1995) do not include types of resemblance that might work for so-called phenomimes or psychomimes in their illustration of communication by virtue of resemblance. However, given the human capability to translate from one sense to the other, it would not be unreasonable to extend the concept of interpretive resemblance to perceptual resemblance. Indeed, as we saw in Chap. 2, expressions such as *ding-dong* show that humans are capable of translating rhythmic movement into rhythmic sounds (Hinton 1995, 4). Perceptual resemblance, as suggested by the name, might not be as verbal as other types of resemblance mentioned by Sperber and Wilson. Most types of resemblance discussed so far are based on linguistic/verbal input. However, relevance theory does not exclude non-verbal communication. As a cognitively grounded theory of communication, relevance theory accounts for all types of communicative phenomena as ostensively communicated stimuli. This means that some phenomena can be used in communication by virtue of perceptual resemblance, rather than interpretive resemblance. The point here is that under this framework, the categories of sound symbolism discussed in Chap. 2, such as the categories proposed by Hinton et al. (1995), *corporeal sound symbolism*, *imitative sound symbolism*, *synaesthetic sound symbolism*, and *conventional sound symbolism*, or the three semantic categories of mimetics (*phonomimes*, *phenomimes*, and *psychomimes*) in Japanese, could all be part of resemblance, whether interpretive or perceptual. Sound symbolism, from this perspective, is part of the human cognitive capacity to entertain representations that range from interpretive to perceptual. So, how does the notion of (perceptual) resemblance enable us to explain the role of onomatopoeia in communication? That is, what does it mean that onomatopoeic expressions are used in communication by virtue of their resemblance to other phenomena? In the remainder of this section, we will review other notions in relevance theory that will help us to answer this question.

1.3 Communication of Impressions

As discussed in Sasamoto and Jackson (2016), onomatopoeia is a highly versatile phenomenon in that some are fully lexicalised examples (such as *sizzle* or *twinkle*), while others are highly novel and not established as a lexical item (such as *gwash*). Similarly, as we have seen in Chap. 1, the meaning of onomatopoeia is often difficult to pin down to a conceptual term, giving rise to nebulous (or ineffable) effects rather than a concrete propositional effect. As such, we need an explanatory framework that accounts for this descriptive effability, as well as for verbal *and* non-verbal communicative behaviour. Relevance theory provides exactly that. First, let us see what *nebulous* or *intangible* means in relevance-theoretic terms.

As Sperber and Wilson (1986/1995) explain, communication is a matter of degree. Not all communicative acts aim to achieve a single, strong cognitive effect. In some cases, some utterances might communicate a single, strongly evidenced proposition that the hearer cannot help but recover. In other cases, utterances might communicate assumptions that only weakly evidence that what is communicated amounts to no more than an impression. In such cases, the recovered range of assumptions is so weak that it is hard to render in propositional terms at all (Blakemore 2008, 2011, 2015; Pilkington 2002; Sperber and Wilson 1986/1995, 2015; Wharton 2009).

Impressions, along with other non-propositional effects such as attitudes and affects, are called expressive meanings and are extremely intangible and almost impossible to spell out in propositional terms (for further discussion, see, for example, Sperber and Wilson 2015; Wharton 2003a, 2009). Sperber and Wilson (2015, 138) define impressions as follows:

> an impression is a change in the manifestness of an array of propositions which all bear on our understanding the same phenomenon, answering the same question, or deciding on the same issue.

Not all propositions that make up the impression might be entertained fully in the interpretation process. That is, the hearer might not process individual propositions *per se*. Instead, as Sperber and Wilson (2015) explain, the hearer will process the array of assumptions together and these assumptions, taken together, lead the hearer to a certain conclusion, giving rise to extremely weak, intangible effects. That is, when what is communicated is so nebulous, the hearer will recover a layer of impressions (or a layer of extremely intangible assumptions), rather than a single, strongly evidenced proposition (i.e. a strong implicature). As claimed in Sasamoto and Jackson (2016), onomatopoeia contributes to the communication of such impressions. That is, what is communicated by the use of onomatopoeia is an array of nebulous and intangible assumptions, which allows for the sharing of impressions.

1.4 Impressions and the *Showing–Saying Continuum*

So, how does onomatopoeia enable the speaker to share impressions? We claimed in Sasamoto and Jackson (2016) that onomatopoeia is located on the continuum of *showing* and *saying*. The idea of the *showing* and *saying* distinction was not first developed in relevance theory. Grice (1957), for example, made a distinction in speakers' meaning between *showing* and *saying*. A typical case of *showing* would involve providing non-verbal evidence, while *saying* typically involves verbal (or linguistic) input. For example, in order to respond to an invitation to play tennis, one might show a bandaged wrist. More often than not, this would be taken as the communicator declining the invitation. This is a typical case of *showing*, where the communicator provides direct (or non-coded) evidence for communication. On the other hand, to the same invitation one might say, "I've injured my wrist and cannot play tennis for another three weeks." This is a typical *saying* case, where the communicator provides indirect, coded evidence for communication. Direct evidence leads the hearer to an intended interpretation on the basis of non-verbal evidence and, as such, requires more inferential steps. In the case of the invitation to play tennis, the communicator is taken as providing direct evidence by presenting their bandaged wrist. On recognising this evidence, the hearer will first draw a conclusion that the communicator has injured his/her wrist and that the injury is severe enough to warrant the use of the bandage. Combining this assumption with existing assumptions about tennis and the use of the wrist, the hearer will arrive at the intended interpretation that one cannot play tennis with an injured wrist. In contrast, indirect evidence helps to recover the intended interpretation on the basis of linguistic encoding and, as a result, requires fewer inferential steps. Processing the communicator's utterance, "I've injured my wrist and cannot play tennis for another three weeks", is enough evidence to draw the conclusion that they cannot play, and does not require further inferential steps.

The biggest difference between Grice's (1957) recognition of the *showing* and *saying* distinction and the relevance-theoretic notion is that relevance theory (see, for example, Sperber and Wilson 1986/1995; Wilson and Wharton 2006; Wharton 2003a, 2009) acknowledges that it is the *showing–saying* continuum, rather than a distinction. That is, the distinction between *showing* and *saying* is not so clear-cut, and there are many cases that exhibit both *showing* and *saying* elements. A typical example would be to say "I am really upset" in a very sad tone of voice, and with a sad face.

Note that these non-verbal cues are not always intentionally used in communication. As first discussed by Grice (1957), it is acknowledged that there is a distinction between *natural* meaning and *non-natural* meaning. Natural meaning is typically delivered by natural signs. For example, a black cloud approaching would often be taken as forthcoming rain. In contrast, non-natural meaning would involve a communicator's intention to communicate, as well as the intention to make clear his/her intention to communicate. However, relevance theory also acknowledges how humans exploit such natural signs in communication (see Sperber and Wilson 1986/1995; Wharton 2003b, 2009). Let us take an example of a bad case of tennis elbow. Seeing how you struggle to twist a corkscrew or pull a face in pain while trying to pick up a heavy bag, an onlooker might recognise that you have some injury and try to help you. This is not a case of ostensive communication: it is a case of accidental transmission of information. You did not pull a face in pain to attract this person's attention, and you did not intend to communicate; you just have a bad elbow. In contrast, you might intentionally use non-verbal cues when you really do not want to carry those bags, or you want the hearer to open the bottle. In such a case, you might exploit such natural behaviour (pulling a face in pain on purpose) to make it clear you *intend* to communicate that you are in pain and want the hearer to take over. In his series of work on non-verbal communication, Wharton (2003a, b, 2009) shows how humans exploit non-verbal cues such as facial expressions, tones of voice, gestures, and bodily movements in ostensive communication. As discussed by Wharton (2003a, b, 2009), exploitation of non-verbal cues in ostensive-inferential communication is a case of overt, intentional *showing* rather than *saying*. So, what does it mean to communicate via the use of non-verbal cues?

Earlier, we saw that according to relevance theory, there are two layers of information when we communicate: the first is the layer of information that the speaker intends to communicate and the second is where the first layer of information is being pointed out intentionally. The difference between *showing* and *saying* lies in the "directness" of the evidence provided for the first layer of information. Cases of *showing* often involve non-verbal and (hence) direct evidence for the first layers of information, whereas cases of *saying* would normally involve providing evidence derived via linguistic decoding (and hence indirect) and, as a result, the interpretation process involves fewer inferential steps, requiring more of an "inferential leap" from evidence to intended interpretation. For example, if a child is told that they cannot watch another episode of their favourite cartoon, they might let their parents/guardian see their frown, a natural sign of disappointment. In this case, the child's disappointment is directly inferable from frowning. This is a case of *showing* direct evidence. Or, the child might provide less direct evidence for the first layer of information by saying "I am disappointed", from which the hearer arrives at the first layer of information by a combination of linguistic decoding and inference. Or, the child might provide both direct and indirect evidence for the first layer of information by saying "I am disappointed" with a very sad, frowning face. In this case, the hearer would arrive at the first layer of information by a combination of direct inference, linguistic decoding, and inference on the basis of linguistic evidence.

The discussion so far has illustrated how relevance theory can account for both verbal and non-verbal communication without treating them independently from each other. Sasamoto and Jackson (2016) argue that onomatopoeia should be handled within a framework that can account for the role of both verbal and non-verbal behaviours in communication. Certain onomatopoeia, such as *sizzle* or *hiss*, are highly lexicalised, whereas others, such as *galumph* or *vaboosh*, are still novel and creative and therefore might not even be seen as words yet. Such a characteristic of onomatopoeia requires a theory that can handle communicative behaviour that is located in the continuum of verbal and non-verbal characteristics, exhibiting similarities with non-verbal communicative behaviours such as facial expressions, bodily expressions, or tone of voice.

As we have seen, in relevance theory, both verbal and non-verbal behaviours are considered equally communicative, and neither takes priority over the other. This is because it is recognised in relevance theory that pragmatic inference plays a central role in communication. That is, pragmatic inference plays a crucial role in interpreting all communicative behaviour, be it verbal or non-verbal. Interpreting any communicative behaviour would involve recovering what the communicator intends to communicate by providing such verbal or non-verbal evidence. As we saw earlier, the hearer, motivated by the presumption of optimal relevance, goes down the comprehension path to determine how communicative behaviours are to be optimally processed, and what type of effects it is relevant to derive in particular contexts. Analysing onomatopoeia in the relevance-theoretic framework enables us to account for the role of onomatopoeia in inference and how its effects are derived in a wide variety of cases.

Furthermore, this framework would allow for analysis of the novel and creative cases of onomatopoeia that have yet to be established as lexical items. This is because relevance theory enables us to account for cases where there is no code or convention of use, and where hence they must be interpreted via inference. As such, Sasamoto and Jackson (2016) argue that onomatopoeia is located on the continuum of *showing* and *saying*, where some are highly established and hence have mainly *saying* elements, while others might have both *showing* and *saying* elements. In some extreme cases, novel and creative onomatopoeia might only involve a *showing* aspect. Note that Sasamoto and Jackson (2016) are not the first to argue that some linguistic forms involve both *showing* and *saying* aspects. In his work on non-verbal communicative behaviour, Wharton (2003a, 2009) discusses the role of interjections in communication in terms of the *showing–saying* continuum, not distinction. In the next section, I will examine his account of interjections, before turning to the discussion of onomatopoeia.

1.5 Interjections and the *Showing–Saying Continuum*

According to Wharton (2003a, 2009), the continuum between *showing* and *saying* works for expressions such as interjections that are generally considered to be at the edge of language and have therefore been regarded as a paralinguistic, or even completely non-linguistic, phenomenon (Wharton 2003a, 175). Intuitively, there are similarities between interjections and onomatopoeia, and both communicate something highly ineffable. Some scholars even suggest that certain cases of onomatopoeia might be analysed as a special subtype of interjection (see Meinard 2015 for a fuller discussion). Indeed, the fact that there are some onomatopoeia that behave as though they are not fully lexicalised suggests that they are located at the edge of language, as with interjections (e.g. Goffman 1981, Trask 1993, cited in Wharton 2003a, 2009). However, syntactically speaking, onomatopoeia and interjections behave in a completely different manner, and there is no reason why they should be treated as the same phenomenon. This suggests that their similarities must lie not in grammar but in what and how they communicate. In other words, what we need to establish is the nature of information communicated by both onomatopoeia and interjections, or, what kind of evidence a speaker could provide for communication by the use of onomatopoeia or interjections. Typically, interjections communicate emotions and attitudes (Wierzbicka 1992; Wharton 2003a; Meinard 2015, 151). Also, the consensus is that onomatopoeia, like interjections, communicate something extremely intangible, indeterminate, and difficult to put in propositional representation, or something highly impressionistic. How is this achieved?

One typical conceptualist approach is that of Wierzbicka (1992). Contrary to prevailing approaches in linguistic semantics, Wierzbicka (1992) maintains that rigorous definitions of words are possible. A typical conceptualist approach to interjections involves presenting a conceptual structure such as (8):

(8) *wow*
I now know something
I wouldn't have thought I would know it
I think: it is very good
(I wouldn't have thought it could be like that)
I feel something because of that.
Wierzbicka (1992, 164)

However, Wharton (2009) shows that the so-called conceptualist approach will inevitably fail to capture the meaning of all interjections, as "*no matter how complex the definition is, there is always a counter argument*" (ibid.: 77, my italics). There are several reasons for this. First, according to Wharton (2009, 76), this approach does not seem to offer an explanation for "the subtle shade of positive meaning that an utterance of *wow* might communicate", when the use of an interjection could communicate a range of states of mind, including surprise, amazement, astonishment, or bewilderment.

The second problem Wharton (2009) raises with the conceptualist approach is that interjections seem to communicate something much vaguer than conceptual structures such as those provided in (8). For example, what *wow* can communicate is so much greater than what can be described conceptually, especially as interjections are often used together with other devices (including prosody). Similarly, the meaning of interjections is highly context dependent and several indexicals have to be assigned to the putative conceptual structure in order to recover the full content of communicated assumptions. In the case of (8), several indexicals (*I*, *now*, *it*, *that*) need to be assigned referents. The fourth problem with the conceptualist approach is the feeling of naturalness. Wharton (2009, 78) argues that "interjections retain an element of naturalness and spontaneity that suggests they fall somewhere *between* the natural and the linguistic" (his emphasis). For example, there is a range of expressions one could employ to express the feeling of pain, including *aaaargh*, which is a natural scream; ouch, which is culture and language specific; and a fully linguistic representation such as *it hurts like hell.* As Wharton (2009) argues, interjections fall between natural and language-specific items, but there is little discussion on such naturalness in studies that take a conceptualist approach. The fifth problem which Wharton (2009) raises with the conceptualist approach is that "intuitions do not support the claim that interjections encode the kind of conceptual structure the conceptualists propose". Indeed, he shows how the replacement of interjections with their conceptual counterpart does not seem to communicate the same feeling:

(9) a. That flamin' hurts! *Ow*!
That flamin' hurts! *I suddenly feel the pain*!
b. I feel pain, I feel pain.
Ow, I feel pain.
c. If I feel pain, I'll tell you.
If *ow*, I'll tell you.

(Examples from Wharton 2009, 79)

Finally, interjections do not contribute to the truth-conditional content of an utterance, while their proposed conceptual counterparts do:

(10) a. I feel pain, the anaesthetic isn't working. If I feel pain, I'll tell you.
b. *Ouch*, the anaesthetic isn't working.

(Examples from Wharton 2009, 79)

According to Wharton (2009), conceptual representations have logical properties and, as such, they can be either true or false, while interjections cannot, as illustrated in (11):

(11) a. Dentist: This won't hurt.
Patient: I feel pain, the anaesthetic isn't working.
Dentist: You're lying.

b. Dentist: This won't hurt.
Patient: Ouch!
Dentist: ?You're lying.

Rejecting the dominant "conceptualist" approach to interjections, and pointing out that the attempt to supply sets of necessary and sufficient conditions for all words generally fails (Fodor et al. 1975, cited in Wharton 2003a: 178),[1] Wharton (2003a, 2009) takes a relevance-theoretic approach in his discussion of what interjections communicate.

Drawing on Kaplan, Wharton (2003a, 2009) argues that interjections have *expressive* meaning where they are typically used to reveal the speaker's emotional states. He draws on the relevance-theoretic distinction between conceptual and procedural encoding and argues that interjections encode procedures rather than concepts. In relevance-theoretic semantics, it is generally acknowledged that there are two types of encoding: procedural and conceptual. Words that encode concepts typically denote what constitutes a propositional representation, such as *sad*, *dog*, or *walk*. In contrast, expressions that encode procedures guide the hearer through the interpretation process, rather than becoming part of the propositional representation that is being processed. Examples of such expressions would include *but* or *nevertheless*. Wharton draws on this distinction and concludes that interjections encode procedures and contribute to relevance by activating representations of a wide range of emotional or attitudinal states.

Earlier, we saw how the *showing* and *saying* distinction is not indeed a distinction, but a continuum. Some communicative behaviours can be used both to *say* and *show* simultaneously. Wharton (2003a, 2009) applies this continuum in his analysis of interjections and argues that interjections can be treated as expressions that *show* and *say* at the same time. Let us examine the case of *damn*:

(12) Ann: Don't forget, you said you'd proofread Kelly's essay tomorrow.
David: Damn!

(Sasamoto and Jackson 2016, 44)

Wharton's analysis predicts that *Damn* in (11) is partly coded, and partly related to something like natural cries (via, for example, the use of

[1] I will revisit the so-called conceptualist approach in Chap. 4, where the lexical aspect of onomatopoeia is discussed.

affective intonation). That is, the *saying* or coded element of *Damn* in (11) activates a representation of his mental state, such as being annoyed at having to correct the essay. In addition to such a *saying* element, the *showing* element of *Damn* provides direct evidence for the communication via a certain intonation pattern, which allows for the *showing* of the mental state.

The analysis of interjections in terms of the *showing–saying* continuum enables us to explain how expressions communicate something as intangible and ineffable as an emotion or an attitude. Considering how difficult it is to convey something so vague and intangible in conceptual terms, it is not surprising that cases of *showing* typically arise when a communicator wishes to communicate emotions, a complex feeling, or even sensory experience. Or, it may be a simple case of "seeing is believing", where providing direct evidence is more convincing than explaining in words, such as showing a bandaged wrist to decline the invitation to play tennis or pointing out black clouds to communicate you think it is going to rain soon. It would, therefore, be reasonable to argue that *showing* allows for the sharing of experiences or draws on experiential elements of the context. In the next section, I will discuss the role of onomatopoeia in terms of the *showing–saying* continuum.

1.6 Onomatopoeia and the *Showing–Saying Continuum*

So far, there has been little research on onomatopoeia in relevance-theoretic pragmatics, or in pragmatics in general. To my knowledge, the only exception is Wharton's (2003a, 2009) mention of onomatopoeia as supporting evidence for his analysis of interjections, and Sasamoto and Jackson's (2016) analysis that draws on Wharton's work. According to Wharton (2009, 99), cases of iconic language use including onomatopoeia involve an interaction between coding and inference processes at the lexical level, where "words", which are arguably linguistic in nature, also seem to exhibit a degree of *showing*; that is, they provide more direct evidence for what the speaker intends to communicate (Sasamoto and Jackson 2016). Thus, in addition to what is encoded, onomatopoeia also *show* (= provide direct evidence for) something about the objects they are used to describe. Wharton (2003a, 2009) predicts that this *showing* is made possible since the link between the sounds of these words and their meanings is not completely arbitrary, as is claimed repeatedly in the sound-symbolism approach. Take the case of *sizzle*, for example. In addition to the concept SIZZLE[2] ("to burn or scorch so as to produce a hissing sound", *OED*), the onomatopoeia *shows* the kind of phenomena that one would experience. This probably is due to the sound (or phonology) of the word, which bears more than a passing resemblance to the phonology of the phenomenon it denotes. That is, the perceptual resemblance between the phenomenon and the sound of the word enables the communicator to *show*, or provide direct evidence for, the phenomenon to the hearer, in addition to communicating what is encoded by the word. Onomatopoeia's nature, which necessarily involves perceptual resemblance, allows for the communication of elements of sensory experience, feelings, or impressions, which are so nebulous and difficult to convey in conceptual terms, unlike purely conceptual expressions such as *octopus* or *star*. Following this, we argue in Sasamoto and Jackson (2016) that the "meanings" of onomatopoeic expressions range from fully established concepts to apparently "non-propositional" effects, and that they do have conceptual semantics in some cases, but that other aspects of their meaning fall more on the *showing* than the *saying* side. That is, in this relevance-theoretic analysis, onomatopoeia is seen as having both *showing* and *saying* elements. Onomatopoeia's *showing* element provides *direct evidence* for communication, bringing about a similar impression/feeling that would be perceived in a different sensory domain. On the other hand, onomatopoeia's *saying* element provides indirect, coded evidence for communication, conveying a specific conceptual meaning, wherever available. Onomatopoeia, therefore, provides quick and dirty access to (extremely weak) non-propositional effects that are based on sensory experience in addition to what is linguistically encoded. Onomatopoeia, in this way, can be seen as humans' exploitation of connections over different sensory domains, allowing for the communication of extremely

[2] It is a convention in relevance-theoretic literature that concepts are represented in small block capitals.

intangible and vague feelings or impressions, in addition to propositional content where such content is available.

Analysing onomatopoeia as involving the *showing–saying* continuum enables us to account for varying degrees of lexicality of onomatopoeia, running from those which are entirely novel and creative, possibly a pure case of *showing*, through a middle range of terms which combine *showing* and *saying*, through to cases which are fully established as a word with little of a *showing* element. Such varying degrees of lexicality have raised concerns for some researchers such as Flyxe (2002) and Kadooka (2005). As reported in Sasamoto and Jackson (2016), these scholars show how the degree of lexicality in onomatopoeic words varies. On the one hand, expressions such as the English *hjckrrh* (Kadooka 2005), which is an exclamation used by Lewis Carroll in *Alice in Wonderland*, or Japanese expressions such as *bakyuun* (Kadooka 2005), which denotes the sound of a gun or *gān* (Flyxe 2002) and is often used when someone is shocked, are among those that are highly novel and the least lexicalised. In the middle of the lexicality scale exist expressions such as *meow* and *pop* in English and Japanese onomatopoeia for dryness, *karari* (Flyxe 2002) and *sowa-sowa-suru* (Kadooka 2005), which denote the state of nervousness in Japanese. On the other end of the spectrum are highly lexicalised expressions that are established as a lexeme. Such expressions include words like the English *chatter* and the Japanese expressions for "surprised", *odoroku* (Kadooka 2005), or "clearly", *kippari* (Flyxe 2002). In our relevance-theoretic framework, the less lexicalised expressions, such as *hjckrrh* or *bakyuun*, would be located at the *showing* end of the continuum, while other expressions that are highly established, such as *odoroku* or *chatter*, would be located at the more *saying* end of the continuum.[3]

This analysis of onomatopoeia based on the relevance-theoretic *showing–saying* continuum is a departure from the sound-symbolism approach, where the iconic feature of onomatopoeia is put down to form/sound and meaning. Instead, this analysis enables us to zoom into cases where onomatopoeia exhibits an interaction between coding and infer-

[3] I will return to the lexicality of onomatopoeia in Chap. 4, where the lexical aspects of onomatopoeia are discussed, and again in Chap. 7, where issues with translating onomatopoeia are discussed.

ence at the lexical level, contributing to lexical-pragmatic processes. According to Wharton (2009, 100), "in onomatopoeic expressions generally, the link between sound and meaning is not *as* loose as in most other words, since some element of the natural connection remains". However, this is not to be taken to mean that there *is* some systematic link between sound and meaning. Instead, the relevance-theoretic analysis of onomatopoeia, in terms of the *showing–saying* continuum, suggests that onomatopoeia's iconic features stem not from the systematic sound–meaning link, but from the way onomatopoeia communicates about what the speaker perceives. If a speaker wants to communicate what they perceive, they should use what they think is a faithful enough representation of it; that is, one with some natural cognitive resemblance to the original perception. In this way, onomatopoeia provides direct evidence for communication. Or, to be more precise, onomatopoeia provides direct evidence for the first layer of information (the intention to inform the audience of something). As such, Sasamoto and Jackson (2016) claim that onomatopoeia communicates via *perceptual resemblance*. That is, what onomatopoeia *shows* is a representation that perceptually resembles the experience the speaker wishes to share. Unlike an interjection, onomatopoeia is not a natural response to stimuli and, as such, is often stylised and iconic. Instead, onomatopoeia is a sound-based tool that the speaker can use to communicate sensory experience, whether or not the original experience is sound based. Onomatopoeia allows for sharing of any sensory experience, be it visual, aural, or by virtue of its perceptual resemblance, since the speaker considers onomatopoeia to be a faithful enough representation of their sensory experience. This suggests that the link between sound and meaning is what *humans* create, rather than being systematic and external to humans' cognitive capacity, as scholars who take the sound-symbolism approach seem to indicate.

In an attempt to share the impression (or report the sensory experience), the speaker "imitates" the experience, using whatever tool available is to them. If onomatopoeia is the best tool to use in a specific context, that is because the expression at that particular time and point enables the speaker to create the most faithful representation of the sensory/perceptual experience. It is not because there *is* a systematic link between sound and meaning. The link between sound and meaning is triggered because the sound is the tool that best communicates what most closely resembles the original experience. As noted in Sasamoto and Jackson (2016), the sound of the onomatopoeia activates certain qualities that are often associated with such sounds (voiceless sound = clarity, for example). However, these qualities are not *encoded* by the sound. Instead, the link between sound and quality is activated only in a certain context and is therefore context dependent.

Earlier, I mentioned how *showing* is typically involved when the speaker wishes to communicate something highly intangible such as feelings and impressions. Recall *jingle* and *tinkle* from Chap. 1 (repeated as (13) below), which illustrate the role of *showing*:

(13) How they *tinkle, tinkle, tinkle,* [...] /From the bells, bells, bells, bells, /Bells, bells, bells— / From the *jingling* and the *tinkling* of the bells.

(Edgar Allan Poe, "The Bells", my italics)

In this example, both *jingle* and *tinkle* express the specific sound of bells. In the *OED*, *tinkle* is defined as *to ring*. The definition of *jingle* is far more complex and includes a comparison with other expressions, as follows: "to give forth a mingling of ringing sounds, as by the striking together of coins, keys, or other small metallic objects; it expresses [...] a more complicated one than *tinkle*." Here, both expressions share the core meaning, which is *to ring*. In addition to this core, established meaning, there is still an element of *showing* involved, which leads to the recovery of different sensory impressions. In the case of *jingle*, the use of voiced consonant /dʒ/ and /g/ as opposed to clear voiceless sound would evoke something more rhythmic than a simple ringing of a bell, as noted in the *OED*, leading to the impression of a more complex sound of more than one bell playing together. On the other hand, a clear sound that consists of *tinkle* evokes less complex, but perhaps more subtle, delicate impressions. That is, although both expressions share a firmly established meaning, there is an extra layer of meaning, which is very difficult to put into words. This is the *showing* element of the onomatopoeia, leading to the communication of a sensory impression as part of the first layer of information that the audience is intended to pick up. That is, as discussed in

Sasamoto and Jackson (2016), using onomatopoeia enables the speaker to communicate a *sensory impression* so nebulous that it is difficult to put into words, in addition to an encoded concept. While well established, lexicalised onomatopoeia may have a stronger *saying* element,[4] their phonetic link to the senses enabling the communicator to include some expressive effects. Let us now see less established onomatopoeia, which seems to evoke a number of different interpretations.

In Chap. 2, I discussed how some onomatopoeia pose a challenge for sound-symbolism-based accounts, as there seems to be a limited systematic link between sound and meaning. Recall (14), a case of a relatively novel onomatopoeia *galumph*:

(14) He went *galumphing* back.
(Lewis Carroll, *Through the Looking-Glass,* my italics)

As discussed in Chap. 2, *galumph* is defined in the *OED* as "to gallop heavily; to bound or move clumsily or noisily". If there is any systematic link between the sound of *galumph* and what it denotes, then *galumph* should only be used for something heavy and unagile. However, the use of *galumph* is perfectly acceptable in the following cases:

(15) A children's nursery rhyme/kid's song
Galumph, went the little green frog one day
Galumph, galumph, galumph, galumph, galumph

(16) We love this big *galumph*, especially when he drinks water. He sticks his whole face in the bowl and comes out smiling with his face dripping
(*Westport Now*, 9 September 2014, my italics)[5]

(17) My feet hurt all night, and I'm constantly *galumph*ing down the street to catch up to my husband who is moving much faster than me because his legs are way longer and also he isn't wearing crazy-town shoes. ("Why I Stopped Wearing High Heels", my italics)[6]

---

[4] This suggests that highly lexicalised onomatopoeia might contribute to other lexical pragmatic processes, such as ad hoc concept construction. I will discuss this in Chap. 4.

[5] http://www.westportnow.com/index.php?/v2_5/comments/meet_lionel/. Accessed 12 May 2019.

[6] https://hellogiggles.com/fashion/stopped-high-heels. Accessed 20 October 2017.

In (15), (16), and (17), *galumph* is used to describe animate objects. In (15) *galumph* is used to denote the movement of a little frog, which would be far from heavy, while in (16) a large mixed breed (mastiff and pit-bull) is called *big galumph*. In (17), *galumph* is used to describe the way a woman walks in high heels. Although one could argue that the onomatopoeia denotes the clumsy way these animals and humans behave, it is difficult to claim that the movement by a frog, the behaviour of a dog, and the manner of the walk of a woman are systematically linked to the same sound, /gəˈlʌmf/, as pointed out in Chap. 2. However, the relevance-theoretic analysis, based on the *showing–saying* continuum, enables us to account for this. While the encoded or the *saying* aspect of *galumph* denotes a certain type of clumsy movement, the *showing* aspect of the expression enables the speaker to share a sensory impression that is difficult to put into words. As argued earlier, the *showing* aspect of onomatopoeia enables the speaker to provide direct evidence for communication, where the speaker uses a word that sounds perceptually similar to the sensory experience the speaker wishes to communicate. The resemblance between the sound and the impression the speaker wishes to share is not systematically linked. Instead, the link is made as a result of a human's inferential processing.

Interestingly, the expression *galumph* is often used in music reviews:

(18) Getting it on still feels a curiously chaste pastime here, thanks to all those keyboard sounds from electronica's infancy, and La Roux's predominantly sweet girlish vocals. She is all for kissing, but not telling ("all I want is to come out of my shell," yearns Kiss And Not Tell), a white funk *galumph* that confirms La Roux and Sherwin have been studying the excellent Tom Tom Club.

(Kitty Empire, "Trouble in Paradise Review—La Roux is Back, on Her Own,"

(*The Guardian*, 20 July 2014, my italics)[7]

(19) What begins as a spry shuffle grows darker as electric guitars *galumph* and grind into the forefront.

(Tom Moon, "First Listen: Adult Jazz, 'Gist Is,'" *NPR Music*, 27 July 2014, my italics)[8]

---

[7] https://www.theguardian.com/music/2014/jul/20/trouble-in-paradise-la-roux-back-synth-pop-review. Accessed 12 May 2019.

[8] https://www.npr.org/2014/07/27/333716267/first-listen-adult-jazz-gist-is. Accessed 12 May 2019.

Example (18) is an extract from a review of an album by synth-pop music act La Roux, while (19) is an extract from a review on the rock band Adult Jazz. Both acts present significantly different styles of music. The music reviewed in (18) is pop and is very catchy in many ways. It is standard pop music, with no surprises. In contrast, the guitar sound of a song described in (19) has a highly complex structure, with an irregular time signature. Listening to both albums back to back, one would not claim that the music of both can be systematically linked to one sound, *galumph*. However, the use of *galumph* communicates how the author of each review perceived their music, rather than what the sound of *galumph means*. In addition to the concept encoded by *galumph*, which relates to a clumsy or heavy movement, the *showing* aspect of the expression enables the author to share the impression as they perceive it. In the case of (18), the onomatopoeia allows for the communication of the impression of a funky rhythm which goes beyond a standard four-on-the-floor groove. On the other hand, for (19), the onomatopoeia enables the author to share the impression of erratic rhythms in the music, led by the guitar sound. In both cases, such impressions are based on the authors' perception of the music, not a systematic meaning, and the use of the onomatopoeia is a way to provide direct evidence, to *show* what the music is like, to the audience.

Examples (14) to (19) illustrate a case where the same onomatopoeia is used to denote a range of phenomena. In contrast, examples (20) to (22) show how the same phenomenon, this time the sound of high-heeled shoes, can be described using different onomatopoeic expressions: *clip-clop* in (20), *tack-tack-tack* in (21), and *click clack* in (22):

(20) Oh, how I longed for the day when I would be allowed to *clip-clop* around in high heels! (Teresa Maria, "Day 10: About High Heel-Obsession And How To Find The Perfect Pair" my italics)[9]

(21) Now we stood, my sister, brother, mother, and I, on the platform of that gloomy, cavernous station, the noise of entering and leaving trains clashing with loudspeaker and cab brakes squealing and the *tack-tack-tack* of high-heeled shoes along the pavement.
(Carl Dawson, *November 1948,* my italics)

[9] https://outlandishblog.com/about-high-heels/. Accessed 12 May 2019.

(22) "All right everyone. I want you to watch Miss Montana" Laforge shouted over the *click clack* of heels returning to their positions. (Gemma Halliday, *Homicide in High Heels, High Heels Mysteries Book #8,* my italics)

Dictionary definitions for these expressions are "imitations of sounds of alternating rhythm" (*OED*) for *clip-clop*, "a combination or pattern of thinner, high click sounds and deeper, more resonant clack sounds" (*OED*) for *click clack*. *Tack-tack-tack*, on the other hand, is not listed as a phrase in itself, seemingly suggesting the more novel nature of the expression. Furthermore, *clip-clop* and *click-clack* share the same consonant cluster, /kl/. *Tack* also contains the consonant /k/ and it seems reasonable to argue that *tack-tack-tack* denotes some repetitive rhythm too. However, from a communicative perspective, what is interesting here is not what each onomatopoeia means. If a rhythmic pattern or a certain consonant alone has a systematic link to high heels, why should there be more than one expression? Instead, these examples show how humans as cognitive beings have a capacity to use different tools to denote the same phenomena and how we, as cognitive beings, can still recover the intended interpretations, despite the non-systematic use of forms. While the *saying* (coded) elements of these expressions deliver a concept associated with an alternative rhythm pattern of a kind, the *showing* elements of onomatopoeia provide direct evidence for communication, allowing us to share impressions via the use of sound that perceptually resembles the sensory experience of the communicator. What matters here is not what each syllable or rhythmic combination in these expressions *means*. The point is that there is no systematic link between sound and meaning, but instead, the communicator uses these onomatopoeia as the best representation available to share sensory impressions. It can even be a one-off expression such as *kik-pik*, as long as it fits what the communicator is trying to communicate in a particular context.

However, it is worth noting here that the prevalence of onomatopoeia with rhythmic patterns has an implication for analysis: it suggests the need to consider the multimodal nature of sensory experience in order to fully account for the role of onomatopoeia in communication:

(23) "Noise of crunchy bones goes *crackety-crack* for miles around."
(Roald Dahl, *The BFG,* my italics)

(24) *Click clack, clicketty clack, clicketty clicketty clack*!
(Benedict Blathwayt, *Night Flight for the Little Red Train,* my italics)

First, these expressions pose a similar challenge to sound-symbolism-based studies, as in the cases of *galumph* where the same onomatopoeia is used to communicate totally different phenomena. The onomatopoeic expressions in (23) and (24) are almost identical, except for the first syllable of each unit being either /kra/ or /klɪ/. However, each expression is used to denote something entirely different. In (23), *crackety-crack* is used to describe the manner in which giants bite into bones, while *clickety-clack* in (24) is used to describe the way in which a little train runs. Would one small difference in a consonant, whether it is /l/ or /r/, makes such a big difference? If one is to take a sound-symbolism approach, that would be the only way to account for the difference. However, as our relevance-theoretic account would suggest, it is the communicator's perception of each sensory experience that dictates the choice of the expression, not the systematic link between sound and meaning.

What is more interesting, however, is their rhythmic combination. Both expressions have a gallop-like structure that consists of a triplet of syllables followed by a single syllable. Both combinations would be interpreted differently from a standard duplication such as *click-click* or *crack-crack*. This indicates that what the communicator wishes to communicate is beyond what is conveyed solely by sound. This rhythmic movement adds another layer to the evidence for communication, where the communicator can *show* what the experience is like. In other words, in addition to the *saying* aspect of each expression (e.g. "A combination or rhythmic succession of thinner, higher click sounds and deeper, more resonant clack sounds", for *clickety-clack, OED*), the use of these onomatopoeia enables the communicator to share the impression of the experience of the little train running or giants munching on bones. That is, by using the multimodal onomatopoeia, which, in this case, consist of sounds and rhythmical patterns that resemble their perception of the sensory experience, the speaker can provide direct evidence for communication, or *show* the experience to the audience.

The next example illustrates this multimodal nature of onomatopoeia more clearly. Recall the case of multimodal onomatopoeia discussed in Chap. 2 (repeated below as example (25) for convenience). This extract is taken from a well-known children's story *Matasaburo of the Wind*, by the famous Japanese author Miyazawa Kenji. Here, a highly creative onomatopoeia *Doddodo dodō do, dodoōdo, dodō* is used to denote a strong wind:

(25) [Describing the strong wind that is storming through the village]
Doddodo dodō do, dodoōdo, dodō aoi kurumi mo fukitobase
MIM blue chestnuts also blow-away
Suppai Karin mo fukitobase *Doddodo dodō do, dodoōdo, dodō*
sour quince also blow-away MIM
"Doddodo dodō do, dodoōdo, dodō; Blow away the green chestnuts too;
Blow away the sour quince too; Doddodo dodō do, dodoōdo, dodō…."
(Miyazawa Kenji, *Kaze no Matasaburo* [Matasaburo of the Wind], 1934, 1.
English translation by Strong and Colligan-Taylor 2002, my italics)

The Japanese language has a number of onomatopoeic expressions for wind, including *pyu-pyu*, *byu-byu*, *hyu-hyu*, *hyuuu*, *sa-*, *soyo-soyo*, and *suu-suu*. These onomatopoeic expressions normally involve the bilabial-plosive or fricative (/s/, /ʃ/, or /h/) consonants and have a duplication of syllables. However, the onomatopoeia in (25), *Doddodo dodō do, dodoōdo, dodo*, is completely different from the standard wind onomatopoeia, consisting of the alveolar plosive /d/ with a complex rhythm structure. This suggests that the impression of the wind described in (25) can only be communicated via the combination of the sound and this specific rhythm, emphasising the multimodal nature of the experience. The voiced alveolar plosive /d/ sound is often associated with heaviness, which allows for recreating the impression of heavy and strong wind, while this particularly complex rhythm communicates the erratic but somewhat musical and melodic movement of the air. That is, the author can exploit the multimodal nature of onomatopoeia that enables them to provide direct evidence for communication, using the link between a particular sound and the perception of the sensory expression. This expression is how the author perceives the experience of wind, and this sensory experience can only be communicated through onomatopoeia that is multimodal in nature. This is not particularly surprising, considering how being in a strong wind is a full-body experience, rather than something experienced only through the auditory channel.

In Chap. 1, I mentioned that onomatopoeia is typically defined as a word that mimics sound. In this chapter, I am arguing that onomatopoeia enables the communicator to share their sensory experience via the perceptual resemblance between the sound of the expression and the sensory experience they wish to communicate. This explanation of onomatopoeia enables us to account for the difference between onomatopoeia and mimetics, and for their shared characteristics. Typically, onomatopoeia is defined as an imitation of a sound, while mimetics are defined as mimicry of non-sound. This suggests that in the strictest sense, onomatopoeia is a case of *showing* in the same sensory domain; that is, showing within the domain of auditory sense. In contrast, mimetics are a case of *showing* across different sensory domains, where *the communicator* uses the sound of an onomatopoeic expression to express a sensory experience from a different sensory domain. That is, the sound of onomatopoeia perceptually resembles the sensory experience from a non-auditory domain, allowing for cross-modal showing. That said, as the examples given illustrate, it is not always possible to isolate one sensory domain over others, where they are linked to a particular onomatopoeia. By defining onomatopoeia in terms not of what it mimics, but of how it mimics (i.e. it mimics the sensory experience via perceptual resemblance, which then is used as direct evidence for communication via *showing*), the distinctions between onomatopoeia and mimetics will no longer be needed, as this can account for a very complex layer of expression in onomatopoeia.

Such cross-modal and multimodal onomatopoeia is nothing special. In English, there are a number of onomatopoeic expressions that have a certain rhythmic structure, such as *ding-dong* or *clap-clap-clap*, where not only the sound itself but also the combination or duplication of sounds is an essential element of what the communicator intends to convey via the use of onomatopoeia.

Recall that in Chap. 2 I showed how Hinton (1995) argues *ding-dong* as a case of movement imitatives, where humans translate "rhythmic movements into sounds, including sound-symbolism forms" (Hinton et al. 1995, 4). In the case of *ding-dong*, the communicator indeed expresses a certain rhythmic movement with sounds. However, it is not just the rhythm but also the sound that *ding-dong* communicates. The analysis of onomatopoeia developed from that presented in Sasamoto and Jackson (2016), in terms of the *showing–saying* continuum via perceptual resemblance, enables us to account for the multimodal nature of onomatopoeia without the need to single out a particular sensory domain. In the case of *ding-dong*, the *saying* element of the onomatopoeia communicates the encoded concept of ringing as a bell/like a bell (*OED*), while the *showing* aspect of the expression communicates an impression of a call–response type of sound of a doorbell via the combination of /ding/ and /dong/, in addition to a heavier sound via the use of the voiced consonant /d/, compared to *tinkle* discussed in (1). Similarly, the duplication of *clap* shows an obliged kind of clapping, with no enthusiasm, in addition to what is encoded by the expression *to make the hard, explosive noise* (*OED*).

While these cross-modal onomatopoeia are highly established, there are cases of novel onomatopoeia that exhibit a cross-modal nature. Sasamoto and Jackson (2016) show how a range of novel cross-modal onomatopoeia are used to denote the agility of objects in a children's picture book. In this picture book, a little girl goes to the field to pick flowers, when various animals such as an elephant and a crocodile fall from the sky:

(25) a. The onomatopoeic expression for a fallen crocodile: *dozuzun*
b. The onomatopoeic expression for a falling elephant: *dokashiin*
c. Onomatopoeic expressions for zebra, a lion, and a panda falling together: *guwashi*, *bako*, *dongorogorogorogoron*

(onomatopoeia used in Izumi Motoshita and Kiyotaka Ishii, *Futtekimashita* [Falling Down] 2007)

In examples (25a) to (25c), onomatopoeia is used to denote large animals falling from the sky. The standard onomatopoeic expression for objects falling and/or banging is *dosu(n)* / *dosa* / *doka(n)* / *doshi(n)* /, /*ban*/ or /*batan*/. And indeed, as these examples show, a /d/ or /b/ sound is used in (25a) to (25c). That is, although none of them has the aforementioned standard form, examples (18a) to (18d) all include /d/ or /b/ sounds, which are often associated with heavy objects, while taking a creative form, and combining these sounds with different rhythmic patterns. *Dokashin* in (25) seems to be made up of *doka* and *doshin*, while *dozuzun* in (25a) seems to be a combination of *dosun* and *zunzun*, which is a somewhat conventionalised onomatopoeia for continuous movement/progression. Furthermore, *dongorogorogorogoron* in (25c) includes *gorogoro*, which is often used to denote a rolling object. *Bako* and *guwashi*, which are possibly related to a collision onomatopoeia *gashi*, denote heavy objects crashing onto the surface. All of these expressions can be seen as cross-modal onomatopoeia, as the communicator uses not only the sound but also the rhythm of the combined syllables to communicate the heaviness and clumsiness of the manner in which these animals are falling from the sky. The use of a novel form of onomatopoeia, rather than the lexicalised onomatopoeia, enables the communicator to share the perception of multimodal sensory experience that goes beyond the auditory domain.

So far, I have analysed examples in which the differences in meaning are relatively obvious. However, differences in what onomatopoeia communicates can be very subtle. Examples (26) to (28) show the use of the Japanese onomatopoeia for silence (Figs. 3.1, 3.2, and 3.3):

Fig. 3.1 Mako Taruishi, *Atatakai Okurimono* [Warm Gift], 1992, my italics

Fig. 3.2 Yuto Tsukuda and Shun Saeki, *Food Wars*, in *Weekly Shonen Jump*, issue no. 16, 15 March 2015, 256. ©YUTO TSUKUDA,SHUN SAEKI/SHUEISHA

(26)

[top right]
Mori no naka wa *shin* to shite kooru yona samusa desu
Forest GEN inside TOP MIM QUO do freeze as-if coldness COP
"It is very quiet and freezing cold in the forest."

(27)

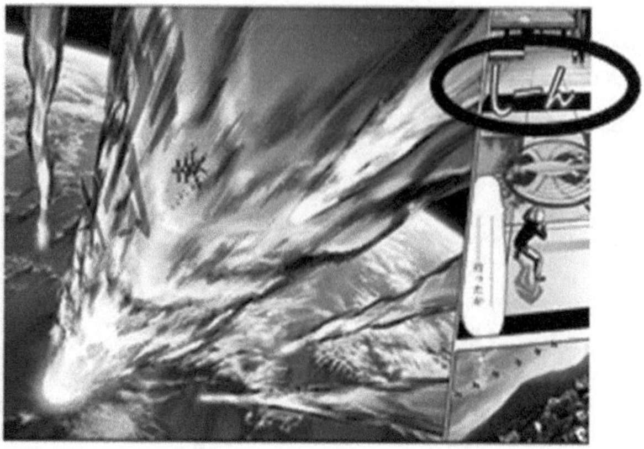

Fig. 3.3 ONE and Yusuke Murata, *One-Punch Man*, vol. 4 2012, 24–25

(28)

Example (26) is taken from a children's picture book. In this scene, animals visit the forest with Christmas gifts for the trees. Here, *shin* is used to communicate the impression of a quiet morning in a winter forest. In (27), a scene taken from a manga, a long-vowelled version *shi-n, as marked in the cell,* is used with "silence" as the English annotation. Here, the characters, who have been gossiping about the protagonist, fall silent when they realise he is within earshot. *Shi-n* is also used in (28) as marked in the top right cell, this time to describe the disappearance of the enemy. What is interesting is not the fact that (variations of) *shin* are used to imitate silence. The point here is that all three situations that are suitable for *shin* communicate different "feels" or impressions of particular silences. The silence in (26) communicates the crisp and peaceful feel of a winter morning, while the silence in (27) involves the awkwardness of the situation. The silence in (28), on the other hand, yields a sense of unknown danger. The point is that the same onomatopoeia communicates a range of different impressions in different contexts, and it is not clear how a sound-symbolism account could explain such subtle differences.

Furthermore, as I mentioned earlier in this chapter, communication is a matter of degree (Sperber and Wilson 1986/1995, 197–202) and not all ostensive communicative acts result in the recovery of a single, strongly evidenced proposition. Instead, there are cases where ostensive communicative acts are only so weakly evidenced that the recovered interpretation amounts only to highly intangible impressions. In such cases, the onus is on the hearer to identify assumptions that satisfy their own expectation of relevance. That is, if communication is only weakly evidenced, then it is likely that some ostensive communicative acts fail to achieve relevance, especially if the hearer does not have enough existing assumptions to draw a conclusion with what has been provided by the communicator. In Chap. 2, I gave an extract from a chat show on Japanese TV, where panellist members with little knowledge of food are unable to follow the discussion (repeated here as example (29)):

(29) Fried rice example (onomatopoeia is marked in italics)

MC: Katsumata (Panellist 1), (how do you like) fried rice?

Panellist 1: I prefer *betabeta*, the sort you get at a greasy spoon. I'm not really into *parapara*.

MC: Oh, you prefer *betabeta*

Panellist 1: (the kind of) fried rice dinner ladies would make

MC: Ah, got it

Panellist 3: But fried rice is not *parapara*. Not *parapara* but *parapara shittori*. (Fried rice) would not <u>taste</u> nice unless (it is) *parapara shittori*.

MC: How about you, Mr Ko (Panellist 3)?

Panellist 3: Current (fried rice) is *parapara* and *shittori*.

MC: Current?

Panellist 3: (it has become) mainstream now

Panellist 2: Once upon a time, *parapara* was good, too, but now it is *parapara shittori*.

MC: Can you make it so?

Panellist 2: Yes

[...]

MC: Do you want to say something, Mr Ozawa (Panellist 4)?

Panellist 4: Well, but fried rice (is/should be) "*bote*"

Panellist 1: Yes!

[laughter]

Panellist 5: He described (fried rice) with sound, "*bote*"!

Panellist 4: You see, you do this (gesture of using a wok) and do *bon* [inaudible]

MC: Can you explain, so that we can understand? (what) "*bote*" [is]? You mean, something heavy?

Panellist 4: Heavy indeed.

Panellist 3: Fried rice should become *fuwa* when you do *pon*

Panellist 5: No, it is not total nonsense, (I can kind of understand) the sound of *bon* when you (flip) the ladle to (serve) fried rice (onto a plate)

MC: How about you, Christine? (Panellist 6)

Panellist 6: I cannot cook at all. I have no idea what you are all talking about, saying things like *parapara* or *betabeta*.

MC: Oh, you do not understand rice being *parapara*, *betabeta* or *don*.

(Extract from "Odoru Sanma Goten," Nihon TV, broadcast 21 June 2016)

As explained previously, the discussion in (29) is centred around fried rice and a number of onomatopoeia are used to describe panellists' favourite texture of fried rice, including *betabeta*, *parapara*, *shittori*, *parapara-shittori*, *bote*, *don*, *fuwa*, and *pon*. Here, while foodie panellists are enthusiastically describing fried rice with a range of onomatopoeia, the other panellists, who have little interest in food, look rather disengaged, and one of them actually claims that she cannot follow the discussion. What is most interesting is the utterance produced by Panellist 6, who is understood as not being able to make the link between the referent—that is, rice—and the texture described by each onomatopoeia. Note that none of these onomatopoeia is novel. All expressions used to describe fried rice in this example are highly established, and it is unlikely that any panellists are unaware of what is encoded by each expression (or the *saying* element of the onomatopoeia). This suggests that it is the *showing* element that Panellist 6 struggles to understand. If this is right, then it would be reasonable to argue that the *showing* or the communication via presenting direct evidence can only be achieved if the hearer is capable of entertaining weakly evidenced assumptions. If they have no experience of the particular texture of fried rice, even if they manage to recover the encoded part of the onomatopoeia, the recovery of impressions, which is communicated via a perceptual resemblance that is *shown* to the hearer, is not possible, unless they have enough existing assumptions.

It is not clear how analyses of onomatopoeia in terms of sound symbolism could account for this type of example, as being "systematic" suggests that a native speaker, if not humans in general, would be able to identify links between sound and meaning. Of course, as acknowledged in the literature, the universality of sound symbolism has yet to be fully explored. However, this example shows that the comprehension of onomatopoeia is not independent of the inferential capacity of individuals.

2 Conclusion

This chapter set out to explore the role of onomatopoeia in communication. To this end, it was argued that sound symbolism alone could not offer an explanation for the context dependency of onomatopoeia. As shown in this chapter, what appears to be a systematic link between sound and meaning is a result of humans' inferential processes. That is, sound symbolism is a repertoire of links that humans are capable of making between sound and meaning, rather than such links existing independently of the human mind. That is, onomatopoeia is humans' attempt to "recreate" what they perceive in their sensory experience, using the tools available to them. Onomatopoeia occurs when such a tool happens to be language. As it is based on perceptual resemblance, the link might seem non-arbitrary, but that is a result of our inferential processes, rather than the language having a non-arbitrary (and hence systematic) link. However, if one insisted that there was indeed an objective, systematic link between sound and meaning external to our cognition, a theory that appeals to such a link should provide an explanation for how a hearer would choose one meaning over another, or how the same sound could be used to denote a range of different phenomena. Furthermore, it is not clear how sound-symbolism approaches can account for deviations from standard onomatopoeic expressions and the creation of new, ad hoc ones. After all, the focus of most works adopting the sound-symbolism approach seems to be on the iconicity between sense and sound, without taking communication into consideration.

In contrast, the relevance-theoretic analysis of onomatopoeia in terms of the *showing–saying* continuum, developed in Sasamoto and Jackson

(2016), provides us with the explanatory force. It enables us to account for humans' cognitive processes, in terms of the relevance-guided comprehension heuristic. That is, the hearer chooses one meaning over others because of considerations of relevance. Or, the hearer can adjust the meaning of one expression in a specific context. We can never pinpoint the "meaning" of a sound. A sound could appear in a wide range of contexts. After all, onomatopoeia is a communicative phenomenon, and it does not matter to the current study whether or not the link between sound and meaning is systematic or non-arbitrary. As such, following Wharton's (2009) original suggestion, the current study, first presented in Sasamoto and Jackson (2016), argues that onomatopoeia is located on the continuum of *showing* and *saying*.

Onomatopoeia exploits the perceptual resemblance between its phonetic form and sensory experience and provides both direct and indirect evidence for the first layer of information that the communicator intends to point out. The *showing* nature of onomatopoeia allows for the communication of nebulous, intangible impressions, which are extremely difficult to put into propositional terms. By using onomatopoeia, the communicator can share with the hearer the impression of their sensory experience, which falls within the *expressive* dimension of communication (Blakemore 2008, 2011, 2015; Wharton 2009). This analysis also showed how the link between sound and meaning in onomatopoeia, which appears to be systematic and non-arbitrary, is in fact a result of human cognitive processing. It is not surprising that onomatopoeia seems to present a non-arbitrary link between sound and meaning, since humans use the combination of sounds that resembles what they perceive, rather than the other way around. This analysis, therefore, enables us to explain how some onomatopoeia appear to be polysemous, or how different onomatopoeia can be used to describe the same phenomenon.

This analysis also has some implications for relevance theory. To the best of my knowledge, most cases of *showing* in the majority of the relevance-theoretic literature are based on bodily expression (e.g. gaze and ostensive sighing, discussed by Wharton 2009), visual evidence (a wide range of examples that provide direct evidence, such as a broken hairdryer or a bottle of aspirin, discussed by Sperber and Wilson 1986/1995), or metalinguistic resemblance (e.g. Sperber and Wilson

1986/1995; Noh 2000; Wilson and Sperber 2012). The analysis of onomatopoeia, in terms of the *showing–saying* continuum, opens up a new possibility for the discussion of *showing* and hence non-verbal communication. Onomatopoeia is unique, as the evidence onomatopoeia provides is neither bodily nor visual, but it is not entirely linguistic, as illustrated in the examples presented in this chapter. Onomatopoeia exploits a perceptual similarity between sounds (i.e. phonological forms) and sensory experiences. The fact that onomatopoeia links our perception and sounds suggests that *showing* by onomatopoeia in a wider sense is a case of cross-modal *showing*, while *showing* by onomatopoeia in the strictest sense is a case of *showing* in the same mode. Furthermore, there has been little discussion in relevance theory in terms of perceptual resemblance, although there is an extensive body of literature for interpretive resemblance. This perhaps reflects the nature of onomatopoeia being located on the continuum from verbal (*saying*) to non-verbal (*showing*). Again, this adds an interesting new dimension to the discussion of resemblance.

Hopefully, this analysis sheds light on how *showing* can take place across different modes and behaviours, and will pave the way for further investigation of the interface between verbal and non-verbal communication.

Bibliography

Blakemore, Diane. 2008. Apposition and Affective Communication. *Language and Literature* 17 (1): 37–57. https://doi.org/10.1177/0963947007085054.

———. 2011. On the Descriptive Ineffability of Expressive Meaning. *Journal of Pragmatics* 43 (14): 3537–3550. https://doi.org/10.1016/j.pragma.2011.08.003.

———. 2015. Slurs and Expletives: A Case against a General Account of Expressive Meaning. *Language Sciences* 52: 22–35.

Carston, Robyn. 2002. *Thoughts and Utterances: The Pragmatics of Explicit Communication*. Oxford: Blackwell http://discovery.ucl.ac.uk/12106/.

Flyxe, Martin. 2002. Translation of Japanese Onomatopoeia Into Swedish (with Focus on Lexicalization). *Africa & Asia* 2: 54–73.

Fodor, J., J. Fodor, and M. Garrett. 1975. The Psychological Unreality of Semantic Representations. *Linguistic Enquiry* 6: 515–531.

Goffman, Erving. 1981. *Forms of Talk*. Oxford: Blackwell. http://www.upenn.edu/pennpress/book/715.html.

Grice, H. Paul. 1957. Meaning. *Philosophical Review* 66 (3): 377–388.

Hinton, Leanne, Johanna Nichols, and John Ohala, eds. 1995. *Sound Symbolism*. Cambridge: Cambridge University Press. https://doi.org/10.1017/CBO9780511751806.001.

Kadooka, Kenichi. 2005. On the Degree of Lexicalization in English Onomatopoeia from a Historical Perspective. *The Ryukoku Journal of Humanities and Sciences* 27 (1): 1–13.

Kita, Sotaro. 1997. Two-Dimensional Semantic Analysis of Japanese Mimetics. *Linguistics* 35 (January): 379–416. https://doi.org/10.1515/ling.1997.35.2.379.

Köhler, Wolfgang. 1929. *Gestalt Psychology*. New York: H. Liveright.

Meinard, Maruszka Eve Marie. 2015. Distinguishing Onomatopoeias from Interjections. *Journal of Pragmatics Complete* 76: 150–168. https://doi.org/10.1016/j.pragma.2014.11.011.

Noh, Eun-Ju. 2000. *Metarepresentation: A Relevance-Theory Approach*. Amsterdam: John Benjamins Publishing.

Ramachandran, Vilayanur, and Edward Hubbard. 2001. Synaesthesia—A Window into Perception, Thought and Language. *Journal of Consciousness Studies* 8 (12): 3–34.

Sasamoto, Ryoko, and Rebecca Jackson. 2016. Onomatopoeia—Showing-Word or Saying-Word? Relevance Theory, Lexis, and the Communication of Impressions. *Lingua* 175: 36–53.

Sperber, Dan, and Deirdre Wilson. 1986/1995. *Relevance: Communication and Cognition*, 2nd ed. Oxford: Blackwell.

———. 2002. Pragmatics, Modularity and Mind-Reading. *Mind & Language* 17 (1–2): 3–23. https://doi.org/10.1111/1468-0017.00186.

———. 2015. Beyond Speaker's Meaning. *Croatian Journal of Philosophy* 15 (2): 117–149.

Toratani, Kiyoko. 2005. A Cognitive Approach to Mimetic Aspect in Japanese. *Annual Meeting of the Berkeley Linguistics Society* 31 (1): 335–346. https://doi.org/10.3765/bls.v31i1.856.

———. 2013. Fukushiteki Onomatope No Tokushusei—Tagisei/Jishosei Kara No Kosatsu. In *Onomatope Kenkyu No Shatei: Chikazuku Oto to Imi*, ed. Kazuko Shinohara and Ryoko Uno, 85–99. Tokyo: Hitsuji Shobo.

Trask, R. Larry. 1993. *A Dictionary of Grammatical Terms in Linguistics*. London: Routledge. https://www.routledge.com/A-Dictionary-of-Grammatical-Terms-in-Linguistics-1st-Edition/Trask/p/book/9780415086288.
Westport Now Media. 2014. Big Lionel Has Found A Home—WestportNow.Com—Westport, Connecticut. *Westport Now*, September 9. https://westportnow.com/index.php?/v2_5/comments/meet_lionel/.
Wharton, Tim. 2003a. Interjections, Language, and the "Showing/Saying" Continuum. *Pragmatics and Cognition* 11 (1): 39–91.
———. 2003b. Natural Pragmatics and Natural Codes. *Mind & Language* 18 (5): 447–477. https://doi.org/10.1111/1468-0017.00237.
———. 2009. *Pragmatics and Non-Verbal Communication*. Cambridge: Cambridge University Press. https://doi.org/10.1017/CBO9780511635649.
Wierzbicka, Anna. 1992. The Semantics of Interjection. *Journal of Pragmatics* 18 (2–3): 159–192. https://doi.org/10.1016/0378-2166(92)90050-L.
Wilson, Deirdre, and Dan Sperber. 2012. *Meaning and Relevance*. Cambridge: Cambridge University Press. https://www.cambridge.org/core/books/meaning-and-relevance/500D9D857CB3274BD23AF6FB381133B7.
Wilson, Deirdre, and Tim Wharton. 2006. Relevance and Prosody. *Journal of Pragmatics, Special Issue: Prosody and Pragmatics* 38 (10): 1559–1579. https://doi.org/10.1016/j.pragma.2005.04.012.

4

Semantics and Pragmatics of Onomatopoeia

1 The Meaning of Onomatopoeia

In Chap. 3, it was argued that onomatopoeia communicates impressions, contributing to the recovery of non-propositional, rather than propositional, effects. It was also argued that onomatopoeia involves both *showing* and *saying* aspects of communication. The *showing* aspect of onomatopoeia involves providing direct evidence for communication based on perceptual resemblance. In this chapter, I turn to the *saying*, or coded, aspect of onomatopoeia, and explore possible links between a communicative account of onomatopoeia and the rich insight into lexical pragmatics that can be gained from the perspective of relevance theory. In particular, I will focus on what has been discussed as the polysemous nature of onomatopoeia (cf. Tsujimura 2001), as in the examples below:

(1a) Doa no totte ga *burabura*-suru.
door Gen knob Nom Mimetic-do
"The doorknob is loose."

R. Sasamoto, *Onomatopoeia and Relevance*, Palgrave Studies in Sound,
https://doi.org/10.1007/978-3-030-26318-8_4

(1b) Asi o *burabura*-si-naide suwarinasai.
legs Acc Mimetic-do-without sit
"Sit without swaying your legs."

(1c) kooen o *burabura*-sita
park Acc Mimetic-did
"I strolled in a relaxed way in the park."

(1d) Otto ga uti de *burabura*-site-iru.
husband Nom home at Mimetic-is doing
"My husband is wasting time at home (without doing anything important)."

(Examples from Tsujimura 2001, 45)

In these examples, the use of an onomatopoeic expression *burabura* appears to communicate a range of concepts, from being loose or not still, as illustrated in (1a) and 1(b); being relaxed, as shown in (1c); and wasting time, as illustrated in (1d). Furthermore, replacing onomatopoeia with fully conceptual counterparts, taken from Ono (2007), does not always communicate the same feeling:

(2) a. *Burabura* depaato o aruki mawatta.
MIM department store ACC walk around
"[SUBJECT] walked burabura around the department store."

b. ??? atemonaku jikan o kakete depaato o aruki mawatta.
Without-purpose time ACC takes department-store ACC walk around
"[SUBJECT] takes time to stroll around the department store without purpose."

Such examples raise a question about what the explicit (or coded) content of onomatopoeia is. In order to answer this question, I will discuss the meaning of onomatopoeia in the context of lexical pragmatics, and demonstrate how the relevance-theoretic approach to lexical pragmatics can provide an explanation for how the intended meaning of some onomatopoeia is recovered, at least for established/highly lexicalised cases. I will argue that what appears to be the polysemous nature of onomatopoeia can be explained in terms of ad hoc concepts, constructed by a narrowing and/or broadening of the encoded concept in a particular context. However, novel and creative onomatopoeia may not encode a clear enough concept upon which ad hoc concepts can be constructed.

This does not mean, though, that the meaning of onomatopoeia is decompositional or that the *showing–saying* aspects can necessarily be separated from one another. In order to establish what onomatopoeia means and how we use onomatopoeia in communication, it is essential to carefully examine both aspects of onomatopoeic meaning.

In this chapter I will focus on Japanese onomatopoeia, as the nature of onomatopoeia/mimetics has been discussed extensively in Japanese linguistics, particularly from the perspective of cognitive linguistics. Research often focuses on the affective nature of onomatopoeia. For example, Kita (1997, 381) argues that "Japanese mimetics are a class of words that are not only referential but also evoke a vivid at-the-scene feeling [...] In most cases, a mimetic evokes some complex combination of sensory inputs and affect, which can be described more accurately as impression than as sensation." He further argues that "Japanese mimetics have a unique psychological effect. They evoke vivid 'images' of an experience, full of affect. This imagery is not only visual, but can also be based on other perceptual modalities and physiological states" (Kita 1997, 386).

While it is generally accepted that Japanese onomatopoeia/mimetics evoke some kind of affective and sensual feeling, it is not entirely clear whether they have encoded meanings at all, and what those could consist of; what they communicate at the explicit level (ad hoc concepts); what the relationship is between the encoded meanings and the feelings evoked; or what onomatopoeia encode or how they evoke such feelings. This is not to deny what has been observed in Kita's (1997) analysis of mimetics. However, further investigation into the meaning of onomatopoeia is necessary to establish what is linguistically encoded and how such encoded meaning contributes to the communication of feelings.

Another aspect that is often discussed is the lexicality and mimeticity of onomatopoeia. It is generally acknowledged, often implicitly, that onomatopoeia are located on a continuum from firmly lexicalised to novel clusters. As I discussed in Chap. 3, some researchers assign degrees of *lexicality* (or lexicalisation) to onomatopoeia, which can be determined based on the possibility of inflection and whether they can occur in a quotation. For example, Kadooka (2005) illustrates degrees of lexicality ranging from the least lexicalised expressions, such as the English *hjckrrh* (an exclamation used by Lewis Carroll in *Alice in Wonderland*) and the Japanese *bakyuun* (sound of a gun); through middle cases, such as *meow* and *pop* in English and *karari* (onomatopoeia for dryness) and *sowa-sowa-suru* (nervous) in Japanese; to the most lexicalised cases such as the English *chatter* and the Japanese *odoroku* (surprised). Flyxe (2002) also lists expressions such as *kippari* (clearly) and *odoroku* as examples of highly lexicalised onomatopoeia, and expressions such as *zabun* (splash) or *gān* (often used when someone is shocked) as among the least lexicalised onomatopoeia. Flyxe (2002) describes the degree of lexicalisation in terms of semantic (Fig. 4.1) and syntactic features (Fig. 4.2).

As shown in Fig. 4.1, Flyxe (2002) considers that the degree of lexicalisation ranges from low to high in terms of the semantic nature of onomatopoeia, where *giongo* (phonomimes) are seen as the least lexicalised, followed by *gitaito* (phenomimes), and *gijogo* (psychomimes) are seen as the most lexicalised. In terms of syntactic features, he considers that the least lexicalised onomatopoeia must be used with a quotative particle *to*, while mid-range onomatopoeia involve manner adverbs and the collocation with *to* becomes optional. Highly lexicalised cases involve adverbs of degree or frequency and nominal adjectives, and the use of *to* is not required.

While Flyxe's (2002) scale might explain the lexicality of onomatopoeia from a linguistic perspective, it is not clear what it means for communication if an expression is highly lexicalised or novel. It is hoped that the discussion in this chapter will shed light on how the lexical aspect of

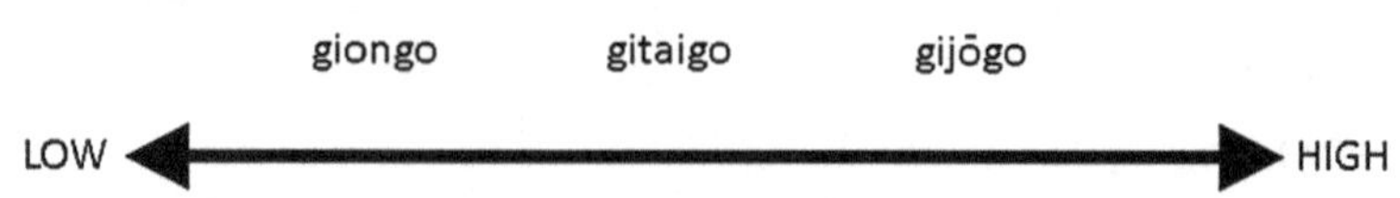

Fig. 4.1 Degree of lexicalisation—semantically (based on Flyxe 2002, 57)

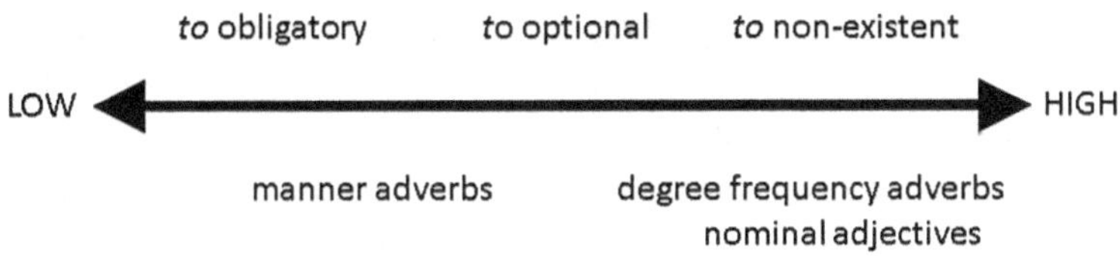

Fig. 4.2 Degree of lexicalisation—syntactically (based on Flyxe 2002, 57)

onomatopoeia is processed in communication and how the degree of lexicality is reflected in the way onomatopoeia is interpreted.

2 Semantic Analysis of Onomatopoeia: Meaning and Sound Symbolism

As Toratani (2018) states, the focus of research on onomatopoeia, especially in Japanese, has mainly been on its phonological and morphosyntactic characteristics (e.g. Martin 1975; Kindaichi 1978; Tamori and Schorup 1999; Akita 2009). However, observations have also been made in relation to onomatopoeia and its status in semantics. As Toratani (2018) points out, the main issues concerning the semantic characteristics of onomatopoeia are its cross-modal nature its polysemous nature where its meaning seems to be "extended", and its lexicalisation and mimeticity. One of the most influential pieces of research on the semantics of onomatopoeia, a relatively new development in onomatopoeia research, is Kita's (1997) two-dimensional analysis.

2.1 Two-Dimensional Analysis of Mimetics

In his analysis of mimetics in Semai (a Mon Khmer language) and Korean, Diffloth (1972) argues that the semantics of mimetics is qualitatively different from that of the rest of the lexicon, and thus proposes a distinction between the analytic and affecto-imagistic dimensions of meaning, where "language has direct contact with sensory, motor, and affective information" (Kita 1997, 380) or where "meaning is represented in terms of affect and various kinds of imagery (auditory, visual, tactile, motoric, etc.)" (Kita 1997, 379). Following this argument, Kita (1997) argues for a separate semantic dimension for onomatopoeia (mimetics) from non-mimetic words, and argues that onomatopoeia works at the affecto-imagistic dimension of meaning, which is distinct from the analytic dimension of meaning, where thoughts and experiences can be analysed in propositional terms. Kita (1997, 387) explains the affecto-imagistic dimension as a dimension where "various kinds of information from different cognitive modalities remain modality-specific, creating the subjective effect of evoking an image or 're-experience'" whereas the analytic dimension is the "'ordinary semantics', where 'meaning is represented as a hierarchical structure of decontextualized semantic primitives'." Kita argues that this dimension is decontextualised in the sense that "it is removed from subjective experience" (Kita 1997, 387).

Kita (1997, 386) further argues that the "semantics of mimetics and that of other parts of a sentence belong to different dimensions" and examines a range of linguistic phenomena which can be attributed to this distinction, including the non-wordiness/semantic redundancy of mimetic adjectives, the fact that they do not fall within the scope of logical negation, their frequent cooccurrence with iconic gestures and prosody, and their iconic morphophonological peculiarities.[1]

This dimension, which Kita (1997) calls the "affecto-imagistic dimension", includes the so-called "expressive function" of language, which scholars often describe as the attitude the speaker shows towards the content of an utterance. However, Kita (1997, 407) also argues that the affecto-imagistic dimension goes beyond what was described as the expressive function, and that the mimetics could also signify a "mental representation of an event of state that is external to a speaker", as well as the affect. Kita (1997) also argues that the affecto-imagistic dimension involves proto-eventuality representation, which carries "the perceptual-motor information which is temporally organized with contingent affective information" (1997, 407). He argues that an affecto-imagistic representation, or a proto-eventuality, is "the minimal inner

[1] Tsujimura (2001) questions Kita's claim based on such evidence. See the next section for a review of her analysis.

code for an experience" (1997, 406). Furthermore, he argues (1997, 406) that "proto-eventualities can also be evoked internally without actual input from the perceptual-motor or affective systems". This suggests that we can "feel" a certain feeli file:///C:/Program%20Files/Adobe/Adobe%20InDesign%20CC%202019/Resources/CEP/extensions/com.adobe.butler.OnBoarding/offline/PropertiesPanel/proppanel.gif ng as if we were indeed experiencing it when, in fact, we are not.[2] An example of this is the way a native speaker feels when they hear a mimetic word being used. For instance, hearing the onomatopoeia *bang* can make one feel as though something had just exploded, even though nothing has. Kita therefore argues that the notion of an affecto-imagistic representation as an eventuality representation explains how mimetics allows for reexperience, saying that "[t]he effect of this evocation is the reexperiencing of the signified eventuality, which leads to the subjective experience of vivid emotive imagery" (1997, 406). This is because, in the affecto-imagistic dimension, various kinds of information from different cognitive modalities remain modality specific, creating the subjective effect of evoking an image or "reexperience" (Kita 1997, 387).

So, at what level does the affecto-imagistic dimension operate? According to Kita (1997, 409), the affecto-imagistic dimension is "the interface between language and other forms of information in mind", where different facets of experience are represented. Such facets might include affective, emotive, and perceptual activation in an experience. However, according to Kita (1997, 387), such facets "do not include the rational construal of it based on such things as agentivity and causality". Kita therefore proposes the affecto-imagistic dimension hypothesis:

> There is a single level of mental representation, the affecto-imagistic dimension, where linguistic information comes into direct contact with sensory, motor and affective information.

Kita (1997, 410) concludes that mimetics should not be treated as a marginal phenomenon, but instead should be considered as "the focus of

[2] Kita is not the first scholar to touch on the distinction between "feeling" a feeling and actually experiencing a feeling. It is often discussed in Japanese linguistics that onomatopoeia allows for the reexperiencing of vicarious experience.

attention in the investigation of functions of language and the language–cognition interface".

In later work, Kita (2001) develops his framework and argues that some onomatopoeia, especially those that are used as a noun, adverb, or verb, belong to the analytic dimension. Similarly, Kita (2013) develops his analysis of mimetics in relation to iconic gestures and argues that the most important characteristics of onomatopoeia/mimetics are how they convey semantic representations. The meaning of onomatopoeia/mimetics is image based, triggered by a quality felt in experience. As a result, onomatopoeia's semantic representation is highly synthetic, since onomatopoeic words encode a wide range of information associated with a certain event/experience. Such an image-based, experience-triggered expression has no arbitrary link between meaning and form, and, as such, even non-native speakers can often infer the meaning to a certain extent (Kita 2013, 79–80). Kita (2013) shows that these characteristics are shared with the iconic gesture, and that onomatopoeia/mimetics and iconic gestures frequently cooccur in natural discourse. Based on this, he claims that both onomatopoeia and iconic gestures share semantic representations and can be used in descriptive rather than indicative communication, due to their image-based semantic representations. That is, mimetics are linguistic representations of images that are encoded based on iconicity through synthetic representation, while iconic gestures are the non-linguistic representation of such a synthetic representation.

While Kita's (1997, 2001, 2013) work seems to capture the intuition about the hybrid nature of onomatopoeia and its meaning, it also raises a few questions. First, it is not entirely clear how having a proto-eventual representation leads to having the experience of vivid, emotive imagery. If the meaning of onomatopoeia is indeed a matter of semantics, as Kita seems to argue, then it would be encoded, and hence it should not be so difficult to pinpoint the meaning of an onomatopoeic expression. Furthermore, it is not clear what he means by "image". This is not to say that Kita's account does not fit our intuitions about the hybrid nature of onomatopoeia. However, crucially, it does not explain what these levels are, nor how these "images" are communicated.

Tsujimura (2001) also points out that mimetic words are not necessarily independent from the analytic dimension. She argues that "[m]imetic words, in and by themselves, are elusive", rather than being independent of other parts of a sentence and singling out a specific manner or subset of semantic participants of the sentence. Drawing on the analysis of mimetic predicates, Tsujimura argues that mimetics are indeed fully integrated with other parts of a sentence, semantically as well as syntactically, and that Japanese mimetic verbs (in the form of mimetics+suru) belong to both the analytic and affecto-imagistic dimensions, raising questions for his dichotomy of two dimensions. As discussed in Chap. 3, relevance theorists have developed a hypothesis that there is a continuum, not a distinction, between fully propositional meaning and non-propositional meaning. While Kita's two-dimensional analysis is not dissimilar to the relevance-theoretic account based on the *showing–saying* continuum, Kita's dichotomic analysis, as Tsujimura (2001) points out, stops short of capturing the nature of onomatopoeia's meaning, which overlaps with what he calls analytic and affecto-imagistic dimensions and therefore does not explain the relationship between the two dimensions.

Similarly, as Toratani (2018) points out, it has also been shown that the linguistic evidence which Kita (1997) attributes to dimensional differences does not necessarily support his argument. For example, as observed by Toratani (2018), Kita's (1997) analysis does not consider mimetic expressions that belong to the analytic dimension, such as nominal mimetics. In addition, his analysis does not account for highly established onomatopoeia that exhibit less mimeticity, such as *tappuri* or *bikkuri*.

Kita's (1997) argument on selectional restriction has also been challenged. Tsujimura (2001) questions his claim that it is the mimetic itself that imposes the selectional restriction for semantic compatibility:

> What is crucial here is that it is misleading to call the type of semantic compatibility [...] "selectional restriction" simply because it is not the mimetic by itself that determines the compatibility between it and agent or between it and theme: in order for the entire sentence to make sense with an adequate modification relation between the mimetic and what is described by it, parts of the sentence need to be considered.
>
> (Tsujimura 2001, 414)

The assumption that underlies Tsujimura's argument against Kita's (1997) view on the semantics of mimetics is that mimetics do not impose any selection restriction independently on the remaining constituents of the sentence in which they are used. Instead, she considers that "mimetics in general have a wide range of reference as a result of metaphorical extension". Indeed, a number of scholars have made an attempt to explain the meaning of onomatopoeia in terms of semantic/metaphorical extension, as I will show in the next section.

2.2 Semantic/Metaphorical Extension and Onomatopoeia

One of the notions often discussed in onomatopoeia/mimetics studies is that of semantic (or metaphorical) extension. Tsujimura (2001) argues:

> And, most importantly, this is because mimetics are essentially used metaphorically and what makes a modification relation "compatible" is determined by the specific metaphorical use that the speaker intends. It is certainly not a property inherent to mimetic words themselves.
>
> (Tsujimura 2001, 414)

This suggests that rather than having a specific meaning, the meaning of mimetics is determined through the interaction with the other constituents of the sentence in which mimetic expressions are used through the process of metaphorical extension. This idea is commonly shared with scholars who examine the meaning of onomatopoeia from the perspective of cognitive linguistics.[3] For example, Inoue (2013) analyses onomatopoeia from that perspective and examines a range of its uses, from the conventional to the novel. Following Kita (1997), Inoue (2013) considers onomatopoeia as belonging to the affecto-imagistic dimension, covering different perceptual modes, including auditory and visual. This cross-modal nature of onomatopoeia suggests that it shares certain characteristics with metonymy, which is based on a part-to-

[3] See Wilson (2011) for a detailed discussion of the parallels in and differences between relevance theory and cognitive linguistics in the treatment of metaphor.

whole relationship (Inoue 2013, 205). She argues that onomatopoeia also involves semantic expansion from the whole to a specific part within the same conceptual domain. For example, the onomatopoeic expression *gatagata* originally denoted the sound of unbalanced items (e.g. the old, rickety chair), but can also be used to refer to non-auditory modes such as vision and touch. Such expansion, according to Inoue (2013), is very much like metonymy, and that is how onomatopoeia becomes polysemous.

The apparently polysemous nature of onomatopoeia is also discussed by scholars working within cognitive grammar. Toratani (2013) examines Japanese onomatopoeia that behave as adverbials, with an aim to identify (i) whether there is a corresponding relationship between the word class of adverbial onomatopoeia and a sense (out of polysemy); and (ii) whether a binary distinction, such as adverbial onomatopoeia representing dynamic events while nominal onomatopoeia represent a static state, can be maintained. Her findings indicate that, in most cases, either one word class denotes more than one meaning (polysemy) or more than one word class denotes one sense. That is, the polysemous nature and word–meaning correspondence of onomatopoeia cannot be determined, nor can it confirm the correspondence between the word class and static/dynamic nature of events that it denotes.

Also, as discussed in Chap. 2, Akita (2013a, c) takes a frame semantics (Fillmore 1982, 1985) approach and examines the semantic extensibility of onomatopoeia. He argues that frame semantics provides a framework account for the links between meanings of an apparent polysemous lexicon and analyses sound-based onomatopoeic expressions from English, Japanese, Korean, and Mandarin Chinese. Akita (2013b) proposes that there are three types of constraint on the semantic extension of onomatopoeia, including referential specificity, event-semantic complexity, and syntactic status. These constraints, as Akita (2013b) argues, determine the semantic extensibility of onomatopoeic expressions.

This discussion shows how what appears to be the polysemous nature of onomatopoeia has been examined in previous studies. While these studies provide insight into the rich repertoire of human cognition and may capture most or all of our intuitions about onomatopoeia, there is little research into *how* the process of semantic/metaphorical extension of onomatopoeia works (or the process of onomatopoeia's semantic/metaphorical extension), or of how the hearer identifies the intended meaning over other possible meanings in a specific context. As Tsujimura (2001) suggests, it may be the case that onomatopoeia (and mimetics) are integrated into purely linguistic structures. However, little has been established as to what is encoded by onomatopoeia and what is pragmatically inferred. Indeed, as Ohori (2018) explains, the focus of studies that take a cognitive linguistic approach, such as those reviewed here, is often to explore the construction of meaning in its totality, without distinguishing semantics from pragmatics. That is, according to Ohori (2018), in these approaches it is not acknowledged that output from the semantic representation feeds into pragmatic inferences. Instead, semantics and pragmatics are seen as being parallel processes (Ohori 2018).[4]

This kind of approach would work if the focus were only on the description of uses of onomatopoeia. However, for an explanatory approach which aims to account for what onomatopoeia communicates and how it is used in communication, it is essential to establish the semantics and pragmatics of onomatopoeia. In fact, as argued in Chap. 3, because onomatopoeia involve perceptual resemblance (or what others have called *iconicity*), they might also be like some interjections in this respect—borderline linguistic (Wharton 2009). This suggests that the meaning of onomatopoeia involves going beyond verbal meaning. However, the studies I have reviewed so far do not precisely address how such a resemblance (or iconicity) is integrated into the semantics of onomatopoeia discussed in the sound-symbolism approach, as the form–meaning relationship.

More significantly, while studies such as those discussed here acknowledge the affective and expressive dimension of onomatopoeia, they often focus only on linguistic aspects (e.g. semantic restriction of agents, verb alterations, negation) and little has been discussed in terms of how

[4] In relevance theory, pragmatics and semantics are also considered as parallel, but not as independent of each other. In relevance theory, the comprehension process is seen as mutual parallel adjustment, in that "hypotheses about explicatures, implicated premises and implicated conclusions are developed in parallel against a background of expectations (or anticipatory hypotheses) which may be revised or elaborated as the utterance unfolds" (Wilson and Sperber 2002, 261–262).

such affective and expressive effects are in fact communicated. That is, focusing on verbal aspects might not necessarily enable us to capture the meaning of onomatopoeia. Furthermore, previous studies often take an etymological approach to the polysemy of onomatopoeia that describes what an expression originally denotes (e.g. *gatagata* originally denotes the sound of unbalanced items such as an old chair) and how the original meaning is semantically extended to refer to non-auditory modes such as vision and touch. That is, these studies focus largely on the contribution of onomatopoeia to semantics, understood as the study of linguistically encoded meanings. As such, while several studies (Kita 1997; Toratani 2005, 2013; Tsujimura 2001) note or imply that there are extra stylistic effects associated with the interpretation of many onomatopoeia, they offer no explanation as to how these effects are recovered. Thus, many of these accounts seemingly concentrate on the relationship between sound and semantics. By contrast, there is no discussion at all of the involvement of pragmatic processes—in particular, the role of pragmatic inferencing—in the interpretation of onomatopoeia. One might argue that there are connections between sounds and interpretations such as a low-pitched sound and heaviness, or a high-pitched sound and sharpness, but merely describing such connections does not present an explanation for how and why these effects are recovered.

To address these gaps in onomatopoeia research, the focus of this study is on how humans take advantage of the perceptual resemblance between onomatopoeia and experience to communicate something highly ineffable. I will propose that this explanatory gap can, in part, be attributed to the lack of a role for pragmatic processes in these accounts. That is, the question should be about the pragmatics of onomatopoeia, not the semantics of onomatopoeia. Indeed, if the meaning (or semantics) of a word is extended, then that would involve certain inferential processes. However, as the studies on onomatopoeia, in terms of semantic extension, have so far been largely focused on the level to which onomatopoeia belongs, little has been said about what onomatopoeia encodes and how the process of semantic/metaphorical extension actually works. This should be discussed in terms of pragmatics.

3 A Relevance-Theoretic Approach to the Meaning of Onomatopoeia

It is generally acknowledged that the meaning of onomatopoeia is ineffable, does not share characteristics of fully fledged conceptual encoding, and is not established to encode concrete propositional meaning. This is because onomatopoeia often communicates sense/feelings/impressions, which are weaker and more intangible than fully fledged conceptual encoding like DOG or RED. However, it does not necessarily mean that it involves procedural encoding that guides the hearer along the comprehension process. Procedural encoding is what it says—it is a matter of encoding. It is stable and, unlike onomatopoeia, does not give rise to nebulous effects. This raises the question of whether the semantics and pragmatics of onomatopoeia can be examined in terms of the standard procedural–conceptual distinction. Wharton (2009) seems to have a similar dilemma, whereby interjections appear to belong to neither conceptual nor procedural encoding. In the next section, I follow Wharton's (2009) analysis of interjections and see if his insight into interjections could be applied to onomatopoeia.

3.1 Interjections and Onomatopoeia: Limits of a Conceptualist Approach

Chapter 3 discussed how Wharton (2009, 89) contested the dominant conceptualist approach to interjections and took a procedural account that addressed most of the problems for which conceptual approaches cannot offer an explanation. As Wharton (2009, 90) argues, interjections activate "various attitudinal concepts or classes of concepts, but not in the standard translational way".

Interestingly, this "standard translational way" (Wharton 2009, 90), in which the conceptualist approach attempts to capture the meaning, has also been taken, explicitly or implicitly, in the description of the meaning of onomatopoeia. For example, Ono (2007, 319) gives the following definitions for the mimetic *burabura*:

a. Describes the motion of a hanging or drooping object swaying under an external force.
b. To stroll about in a relaxed way.
c. To live one's life or pass one's time idly without any particular aim.

While it seems reasonable to assume that onomatopoeia does not fall within procedural encoding, this kind of conceptualist approach still does not capture the characteristics of onomatopoeia. Nor does it explain the vague impression communicated by the use of onomatopoeia. Furthermore, replacing *burabura* with its conceptual counterpart *atemonaku jikan o kakete aruku sama* (the manner in which one walks aimlessly for a long time; Ono 2007, 404) does not communicate the same feeling:

(3) a. *Burabura* depaato o aruki mawatta.
MIM department store ACC walk around
"[SUBJECT] walked burabura around the department store."

b. ??? atemonaku jikan o kakete depaato o aruki mawatta.
Without-purpose time ACC takes department-store ACC walk around
"[SUBJECT] takes time to stroll around the department store without purpose."

While the use of *burabura* in (3a) communicates that the speaker simply spends some time shopping in the department store, its putative conceptual counterpart in (3b) communicates that the person in question is somewhat emotionally lost.

Most of all, the conceptualist account does not explain the range of meanings that onomatopoeia seems to communicate. As previously mentioned, Ono (2007) proposes that there are three conceptual structures of onomatopoeia. However, *burabura* can be used to describe a much wider range of objects and states. Let us see how the use of *burabura* could communicate a range of different impressions depending on context:

(4) a. [a busy mum has a rare day off. Asked what her plan is, she happily declares:]
Machi de *burabura* shiteru yo.
Town LOC MIM do-ing SF
"[I] will be taking some time for myself in town."

b. [Being asked about your daughter, who has never worked and has relied on your money]
Machi de *burabura* shiteru yo.
Town LOC MIM do-ing SF
"[She] will be idling away in town."

c. [Responding to someone whom you are waiting. You know the person is sorry to keep you waiting, so you do not want her to feel bad]
Machi de *burabura* shiteru yo.
Town LOC MIM do-ing SF
"[I] will be passing the time in town."

While all utterances in (4) communicate that the subject will either *stroll about in a relaxed way* (definition b) or *pass their time idly without any particular aim* (definition c), the utterance with the same linguistic representation communicates somewhat different impressions in different contexts. In the case of (4a), the use of *burabura* communicates something pleasant, while *burabura* in (4b) communicates an undesirable, idle way of spending one's life. *Burabura* in (4c) also communicates something pleasant, but not to the extent of (4a). A conceptualist account of onomatopoeia would need to be able to explain such subtle differences. So far, such an account fails to capture the differences. In order to explore the meaning of onomatopoeia, it is essential that we understand the property of onomatopoeia and what it lexically encodes. That is, we need to figure out the division of labour—what is encoded by onomatopoeia (i.e. the semantics of onomatopoeia) and what aspect of the meaning of onomatopoeia is beyond linguistically encoded meaning (i.e. the pragmatically inferred meaning of onomatopoeia).

As I have argued, onomatopoeia does not exhibit a fully fledged conceptual encoding. This means that while onomatopoeia seems to fall within the category of conceptual encoding, as it is truth-conditional and is part of linguistic representations, it still is not fully conceptual as it is (i) often ambiguous and difficult to spell out in conceptual terms; (ii) difficult to pin down in terms of its meaning, which is very vague; (iii) contextually dependent; (iv) is based on natural/iconic meaning with a role played by resemblance in the recovery of that meaning; and (v) expressive rather than descriptive.

This is similar to Wharton's (2009) argument for interjections. Wharton (2009) states that the question of an interjection's procedural encoding remains open but the evidence seems to suggest that interjections work as a pointer to activate a representation of a certain mental state. Drawing on Wharton (2009), in Chap. 3, it was argued that onomatopoeia contributes to relevance by providing both direct (uncoded) and indirect (coded) evidence to communication. That is, the use of onomatopoeia communicates "feeling" via the *showing* aspect of communication, or by providing direct evidence for the speaker's intention. As also discussed elsewhere (Rohan et al. 2018; Sasamoto et al. 2017; Sasamoto and Jackson 2016), direct evidence for communication which the use of onomatopoeia provides is based on the perceptual resemblance between the linguistic form of onomatopoeia and the sensory experience the speaker wishes to communicate. The use of onomatopoeia brings about a similar "impression/feeling" that is perceived in a different sensory domain. Onomatopoeia, therefore, provides quick and dirty access to (extremely weak) non-propositional effects that are attributed to the sensory experience it perceptually resembles. Interpreting onomatopoeia involves a complex process that is guided by considerations of relevance, rather than being a simple case of matching between a sound and a sense from another domain. Onomatopoeia *shows* that what the speaker thinks resembles the experience, using a certain phonological form. That is, onomatopoeia involves a resemblance between the sound and the experience the speaker wants to share.

Furthermore, as discussed throughout the current study as well as elsewhere (cf. Sasamoto and Jackson 2016; Rohan et al. 2018), onomatopoeia also has a *saying* element, as it is located on the continuum of *meaning* and *showing*, providing both coded and uncoded evidence for the speaker's intention. That is, onomatopoeia provides a cue to ostension, using both direct and indirect evidence, thus allowing for activation of the representation of the experience which the speaker intends to share. However, if onomatopoeia provides both direct and indirect evidence, and direct evidence is based on the perceptual resemblance between the linguistic form and the sensory experience the speaker intends to communicate, what indirect (or coded) evidence does it communicate? This chapter focuses particularly on the apparent polysemous nature and elusiveness of meaning of onomatopoeia. In the following section, I will turn to notions developed within lexical pragmatics, one of the areas within relevance theory that focuses on word meaning in context (cf. Carston 1997, 2002, 2010; Sperber and Wilson 1998; Wilson 2003, 2004; Wilson and Carston 2006, 2007).

3.2 Relevance Theory and Lexical Pragmatics

In her discussion of words that encode concepts (i.e. not procedure), Wilson (2003: 344) explains what words encode, as follows:

> mentally-represented concepts, elements of a conceptual representation system or "language of thought," which constitute their linguistic meanings and determine what might be called their linguistically-specified denotations.

As Carston (2010, 9) further explains, such mentally represented concepts are encoded by words and function as an address to recover the intended concept:

> [t]his conceptual address (or file name) gives access to a repository of mentally represented information about the concept's denotation, some of which is general and some of which, such as stereotypes, applies only to particular subsets of the denotation.

However, it is generally acknowledged that words and communicated concepts do not necessarily correspond with each other. Therefore, lexical pragmatics is concerned with "the processes by which linguistically-specified ('literal') word meanings are modified in use" (Wilson 2003, 343). Unlike Grice's approach to conversational implicature, it is acknowledged in relevance theory that pragmatic inference is involved even at the level of identifying the conceptual content of the utterance. The underlying assumption is that what is encoded by a word often differs from what is communicated by the word. That is, "the contribution of words with conceptual meanings involves more than simply accessing the concepts encoded by the words and slotting them into semantic representations" (Clark 2013, 240). As discussed extensively in the relevance-theoretic literature (Carston 1997, 2002, 2010; Sperber and Wilson 1998; Wilson 2003, 2004; Wilson and Carston, 2006, 2007; Sperber and Wilson 2006, to name but a few), the encoded concepts are often modified so that the content communicated reflects "specific meanings intended by communicators" (Clark 2013, 217). In this view, inference is involved not only at the level of recovering implicatures, but also at the level of recovering the concepts encoded by each lexical item. The hearer, following the relevance-theoretic comprehension procedure, uses the linguistically decoded meaning as a first step to reach the intended concept. This process is guided by considerations of relevance. Example (5) illustrates this:

(5) Alfie is happy.

The utterance interpretation process for (5) involves an inference that determines the degree of happiness. For example, if this utterance is produced in the middle of Alfie's birthday party, what *happy* (ecstatic after being given a new bike, for example) communicates would be something much stronger than if this was said as a general description of the child, who is typically pretty laid back and complains very little. Alternatively, if this was said at a business meeting by someone referring to an executive called Alfie, it may be the case that Alfie does not have any objections to the proposal that has just been put on the table. That is, the utterance "Alfie is happy" licenses different interpretations, depending on the context in which the utterance is produced. *Happy* contributes to different conceptual content in each context. Such modification of the meaning of *happy* is guided by considerations of relevance.

The processes involved in such modification include narrowing, where "a word is used to convey a more specific sense than the encoded one, resulting in a restriction of the linguistically-specified denotation" (Wilson 2003, 344), or broadening, where "a word is used to convey a more general sense, with consequent widening of the linguistically specified denotation." As Wilson (2003) shows, broadening also involves category extension, such as using salient brand names or proper nouns to denote a broader category.

(6) All doctors *drink*. (narrowing)
(Wilson 2003, 344)
(7) This coat cost *1,000* dollars. (approximation/broadening)
(Wilson 2003, 345)
(8) We need to get a new hoover. (category extension/broadening)

The concept communicated by the use of *drink* in (6) is not just drinking any liquid. Instead, a more specific concept "alcoholic drink" is communicated. That is, the concept of *drink* is made more specific via the process of narrowing. In (7), the price of the coat is rounded up to *1,000* dollars, rather than a more precise number (say, 999 dollars 99 cents). Similarly, in (8), the word *hoover* is used to denote items in a broader category, not just the vacuum cleaner made by the brand *Hoover*.

Previously, these processes have often been analysed separately as distinct processes (e.g. Wilson 2004). However, as Wilson explains, these are not two distinct processes. Whether it is a case of narrowing or broadening, the encoded concept is used as an input into inference processes. That is, in both cases, the encoded concept provides a blueprint for the inference and the hearer develops this blueprint to recover the intended concept that achieves optimal relevance. Thus, the word is a tool to activate the particular concept in the hearer's mind, in order to recover an intended concept that would satisfy the search for optimal relevance. The intended concept is recovered as a result of inference in context based on existing assumptions. These ad hoc concepts are constructed as one-off conceptual elements in a particular context, used to represent an entity that is not encoded but accessible in our conceptual system (Carston 2010, 15). Ad hoc concepts can be broader or narrower than the encoded concept, which is no more than a clue to help the hearer to construct the ad hoc concept that is suitable in a specific context and satisfies relevance.

The examples below, taken from Carston (2002), illustrate this:

(9) a. There is a large square of lawn at the back.
b. This steak is raw.
c. Ken's a (real) bachelor. [where Ken is legally married]

In (9a), uttering *square* triggers the relevance-driven ad hoc concept formation, as it is doubtful that four sides of the area in someone's garden have precisely the same length at right angles to each other. Instead, the area would be *approximately* square, where it is reasonable to call it SQUARE*. That is, it is called "square" to communicate the concept SQUARE*. Similarly, in (9b), it is very unlikely that one would be served completely uncooked meat. Instead, *raw* in (9b) communicates that the steak is much less well done than the speaker had expected. In the case of (9c), if the person is married, he cannot be a bachelor by definition. Instead, in this case the encoded meaning of *bachelor* is broadened to mean BACHELOR* to describe someone who exhibits a number of traits which can be considered typical of unmarried men. According to Carston (2002, 84), in each case the communicated concept, which is "the extension of the pragmatically inferred concept*", is broader than the lexical concept and includes certain respects that are not in the lexical concept.

At this stage, it is important to note that these pragmatic processes are a continuum. Speaking of literal and non-literal meaning, Clark (2013) explains how relevance theorists take what he calls "the continuity hypothesis". According to Clark (2013, 251), "literalness is a matter of degree and utterances may be more or less literal", and non-literal utterances such as loose, hyperbolic, and metaphorical utterances require no special processes, as distinct from those involved in understanding literal utterances. This is particularly important when it comes to explaining the apparently polysemous nature of onomatopoeia, as it allows for the unified account of the meaning of onomatopoeia and polysemy as a result of pragmatic inferencing, rather than the view that some onomatopoeia is polysemous, as claimed in previous studies.

3.3 Onomatopoeia and Lexical Pragmatics

So far, I have reviewed fundamental notions in relevance-theoretic lexical pragmatics. Next, drawing on these notions, I will argue that onomatopoeia can be interpreted according to the same pragmatic processes as discussed above. The coded element of onomatopoeia triggers construction of an ad hoc concept in a specific context, while direct evidence provided by the use of onomatopoeia adds further impressions beyond lexically encoded meaning. A more lexicalised onomatopoeia would trigger the construction of ad hoc concepts, while what is communicated by the use of less lexicalised, more mimetic expressions might not amount to a fully fledged ad hoc concept and might amount only to ineffable feelings or impressions, as discussed in Chap. 3. The implication of this is that the so-called semantic extension of onomatopoeia, in fact, is a case of the continuum from a literal to a less specified use of onomatopoeia, to reflect the intended meaning. This analysis will show how the notions from lexical pragmatics will enable us to provide a unified account of the polysemous nature of onomatopoeia. In the remainder of this chapter, I will examine a range of examples in terms of the coded evidence provided by the use of onomatopoeia.

The apparently polysemous nature of onomatopoeia is well documented (cf. Toratani 2018; Mikami 2004; Trsujimura 2001). Example (10) illustrates this clearly:

(10) a. kare wa *dondon* to doa o tataita.
he TOP MIM QUO door ACC hit
"He banged the door."
b. Iroirona koto ni *dondon* chosen-shite hoshii.
various matter to MIM challenge-do want
"I'd like you to take on a challenge up different things."

In (10), the same sound *don*, for example, could be used to represent entirely different ways of performing different actions. While *dondon* in (10a) is a stylised imitation of the sound of banging a door, in (10b) it is used to denote a more abstract concept: how a person deals with life. In both cases, *dondon* would have a common characteristic: the manner or sound of hitting something with force repeatedly. Let us say that this is the encoded concept of *dondon* (DONDON). The use of *dondon* in (10a) seems to be somewhat literal and is highly mimetic. In contrast, the encoded concept DONDON is adjusted to reflect something more specific and suitable in this context to mean DONDON* to describe the lively manner in which the person goes head on with life's challenges that come their way.

Examples (10) above and (11) below demonstrate similar context dependency of the interpretation of onomatopoeia. Tsujimura (2001, 45) lists how *burabura* seems to convey a range of meanings, as seen in (1), and Mikami (2004, 3) reports similar elusiveness for *gorogoro*, as shown in (11):

(11) a. Doa no totte ga *burabura*-suru.
door GEN knob SUB MIM-do
"The doorknob is loose."

b. [teacher to a pupil]
Ashi o *burabura*-si-naide suwarinasai.
legs ACC MIM-do-without sit
"Sit without swaying your legs."

c. kooen o *burabura*-sita
park ACC MIM-did
"I strolled in a relaxed way in the park."

d. Otto ga uchi de *burabura*-site iru.
husband SUB home LOC MIM-is doing
"My husband is wasting time at home (without doing anything important)."

What is common among the uses of *burabura* in (11a–d) is the manner in which something (or someone) is not firmly fixed and is at a loose end. Let us say this is the coded element of *burabura* (BURABURA). The use of *burabura* in (11a) would be the closest to this concept. The encoded concept BURABURA is adjusted in each context, so that the recovered concept is more specific and suitable to communicate the meaning intended by the speaker. For example, in (11b), BURABURA is adjusted to mean BURABURA*, which would involve the lack of focus and the mindless manner of the student in the class. Unlike (11a) and (11b), the use of *burabura* in (11c) and (11d) is much broader than the literal use and is metaphorical in that in both cases, it communicates something broader than its literal sense; that is, something that is not fixed firmly means something carefree or liberated. However, at the same time, the expression is modified in each context to mean something more specific. For example, the use of *burabura* in (11c) would communicate the pleasant and carefree nature of the time the speaker is having in a park (let us say, BURABURA**). In contrast, *burabura* in (11d) involves certain characteristics of being "carefree", but to the extent that it is seen as irresponsible or lazy (let us call it BURABURA***).

Earlier, I mentioned that Ono (2007, 404, my translation) treats *burabura* as polysemous and presents the following three definitions:

(12) a. Describes the motion of a hanging or drooping object swaying under an external force.
b. To stroll about in a relaxed way.
c. To live one's life or pass one's time idly without any particular aim.

While such a polysemous account might describe the possible uses of the expression, it is not clear how the hearer chooses one interpretation over another, or in fact, how these three meanings are related. In contrast, the above account in terms of broadening/narrowing and ad hoc concept formation would enable us to explain how and why *burabura* can be interpreted as definition (a) in one context but as definition (b) or (c) in another.

Next, *gorogoro* is also seemingly polysemous at first glance:

(13) a. [The sound of thunder, or a thunder-like sound]
Enrai ga *gorogoro*-to narinagara dandan chikaduite kuru youda.
far-thunder SUB MIM-QUO roaring dandan approach-come looks-like.
"It looks like thunder is gradually approaching."

b. [The manner in which heavy objects or bodies roll in sequence]
danborubako o katamukeru to, migotona jagaimo ga *gorogoro* korogarideta.
cardboard box ACC tilt case impressive potatoes SUB MIM roll-out PAST
"When we tipped the cardboard box, impressive-looking potatoes came rolling out."

c. [To spend time without working or doing anything particular]
shisshoku shite inaka no oyamoto ni kaeri, ichinen hodo *gorogoro*-to kurashiteita.
lost-job did hometown GEN parents to return, a year about MIM-QUO lived.
"(I) lost my job, went back to my parents, and lived doing nothing for about a year."

d. [The way an item does not have a unique quality]
ano teido no bijin nara, Tokyo ja *gorogoro* iru yo.
That degree GEN beauty if, Tokyo in MIM exist SF.
"These beauties are nothing special in Tokyo."

e. [The feeling of discomfort when a foreign body enters the eye]
Gomi ga haitte me ga *gorogoro* suru.
Rubbish SUB enter eye SUB MIM do.
"Something got into my eyes, and it hurts."
(Examples Atoda and Hoshino (1995), cited in Mikami (2004), *Giongo Gitaigo Shihoho Jiten* [Dictionary of Phonomimes and Phenomimes Use]

Gorogoro is an aural-based onomatopoeia that is a stylised imitation of the sound of something rolling or tumbling down, as illustrated in (13a) and (13b). This onomatopoeic expression has become lexicalised enough to appear in dictionaries, and it is now used to describe something more abstract, as seen in (13c) to (13e). However, all these uses communicate a certain characteristic of items rolling or scattered around. It would be reasonable to say that *gorogoro* in (13) was originally purely mimetic, as in (13a), and initially involved little of a *saying* (or coded) element.

However, such a mimetic (or *showing*) element has become lexicalised and has come to mean GOROGORO, which has now become pretty stable as the encoded concept. Even when *gorogoro* has become stable, the expression is still context dependent, and the notion of ad hoc concepts explains how subtle differences are communicated in each context. For example, (13b) is slightly different from other examples, as the event also involves physical objects. In this case, GOROGORO is broadened and means something specific, GOROGORO*, which describes not just the sound but also the manner in which large potatoes come rolling out of the box (category extension). In (13c), what is communicated by the use of *gorogoro* is even more abstract, and GOROGORO is modified to describe not the sound, but a person's behaviour/state of mind. Let us call this GOROGORO**. GOROGORO** is broader than the lexical concept GOROGORO and shares certain characteristics with it. In this case, there is no sound involved. Instead, it is an extension of the encoded concept GOROGORO and communicates a certain state of a person's life where they are lazing around. *Gorogoro* in (13d) also communicates an extension of the lexical concept GOROGORO. Here, the intended concept of *gorogoro* (say GOROGORO***) is also broader than the encoded concept and refers to the state where objects are scattered all over the place as a result of rolling out of their original place. Finally, in (13e), *gorogoro* communicates GOROGORO****, which describes the way something is in the eye causing the discomfort.

At this stage, it is important to note that these ad hoc concepts are not necessarily paraphrasable. Carston (2010, 165) argues that "ad hoc concepts are, generally, ineffable, in the sense that, as well as not being lexicalised, there isn't a linguistic phrase that fully encodes them either, and the paraphrases are intended as just a rough indication to aid readers in understanding what we have in mind in particular cases". This, again, supports the idea that the coded aspect and its seemingly polysemous nature can be explained in terms of pragmatic adjustment of lexical concepts and formation of ad hoc concepts.

The examples I have examined so far only involve fairly standardised and highly lexicalised onomatopoeia. How about less lexicalised, creative (or novel) onomatopoeia? The advantage of the lexical modification account, over the polysemous account, is that it enables us to account for even novel examples such as (14) and (15):

(14) がん/がんがん/があーん (gan//gangan/gaan)

a. *gan* to iu oto ga shita.
MIM QUO say sound SUB did
"There was a sound of *gan*."

b. Atama ga *gangan* suru
head SUB MIM do
"I have a thumping headache."

c. Sono news o kiite *gaan* to shokku o uketa.
That news ACC heard MIM QUO shock ACC received
'I got shocked by hearing the news."

As example (14) illustrates, the onomatopoeia *gan* takes various forms. In the sound-symbolism approach, it is acknowledged that the variety of form is also an indication of sound symbolism. For example, some authors, such as Hinton et al. (1995), argue that duplication of a syllable means the repeated nature of the action. While it is likely that repeated use of the same or alternate syllables could *communicate* such a repeated nature of the action, what is not explained is how duplication could have a systematic link with (or encode) repeated actions. The notions from relevance-theoretic lexical pragmatics enable us to account for such phenomena.

Gan in (14a) imitates the sound of hitting a hard surface using a metal object. As Moriyama (2007, 57) explains, this onomatopoeia is highly established and has been in use since the fourteenth century. *Gan* in (14a) is the typical use where the expression is employed to describe such heavy sounds. It would be reasonable to assume that it encodes the lexical concept GAN, which is established, or lexicalised, as a result of the process of lexicalisation from being a pure onomatopoeia based on the sound. This lexicalised concept is adjusted to suit a specific context. For example, as illustrated in (14b), the expression can also be used to describe pain or a state of shock. In (14b), it is used as a duplicated form and expresses a thumping headache. The denotation of the expressed concept GANGAN* is broader than GAN, as the category is now broadened to include pain that is not caused by contact with a heavy object. The duplication of the form *gan* provides a cue for the hearer that the pain is not one-off and is repetitive. This account is different to a sound-symbolism approach in that, rather than claiming that duplication has a systematic link with the repeated nature of an action, our claim is that the duplication contributes to the formation of an ad hoc concept, by providing a cue to the construction of this concept via its phonological form and the duplication of the phonological form. It is the humans that link the duplication of the form with the repetition of an action and the onomatopoeia provides the cue for the interpretation. Finally, in (14c), *gaan* is used to describe a state of shock. While the use of *gaan* in this way is not necessarily novel, it has been somewhat confined in the colloquial use and is acknowledged in dictionaries only in passing. For example, while Moriyama (in Ono 2007, 57) spends a whole page describing the meaning of *kan*/*gan*, which is the root form of *gaan*, there is only one sentence saying that "*gaan* also represents the state of shock" (Ono 2007, 57). So, how is this particular use of *gaan* interpreted? The concept expressed by *gaan* (GAAN*) is certainly broader than GAN, as the category is broadened to incorporate the state of shock, which does not even include any pain that would be caused by the contact with a hard object. Instead, it is broadened to denote the aftershock of being hit by something emotionally traumatic. The lengthening of the vowel leads to the construction of the ad hoc concept, as the phonological form provides a cue to the hearer that there is a prolonging effect of the trauma.

Furthermore, as shown in example (15), there are cases of compound onomatopoeia:

(15) a. [Description of bread]
shittori ama*fuwa* ren-nyu pan
moist sweet-MIM milk-bread

"moist, light, sweet milky bread."
(Nangumo, "Shittori Amafuwa Rennyu-Pan," 2014)[5]

b. [Caption for a model photo]
ama*fuwa* girlie waibu
sweet-MIM girlie wave

"sweet and airy girlie curly-hair."
(beauty.hotpepper.jp, no date)[6]

[5] https://cookpad.com/recipe/2684995 Accessed 12 May 2019.
[6] https://beauty.hotpepper.jp/slnH000231469/style/L003653319.html. Accessed 12 May 2019.

In (15a) and (15b), the onomatopoeia *fuwa* is combined with *ama*, which is part of the adjective *amai* (sweet). This kind of composition is entirely context dependent, and not all examples are established as a lexeme. Instead, these are examples of hybrid concepts via lexical modification. In both (15a) and (15b), this semantically complex expression denotes something airy and sweet, and the composition of *ama* and *fuwa* guides the hearer to combine the denotation of AMAI and FUWAFUWA, which leads to the formation of an ad hoc concept AMAFUWA*. However, this does not mean that *amafuwa* in both examples communicates the ad hoc concept AMAFUWA*. The lexical concepts AMAI and FUWAFUWA are modified (broadened and/or narrowed) and compounded to suit the context. In (15a), where the expression denotes the quality of the bread, the lexical concept AMAI is narrowed so that it only includes certain sweetness (and excludes sickly sweetness such as treacle), while the lexical concept FUWAFUWA is modified to FUWAFUWA*, which denotes a specific airiness suitable for bread, excluding the lightness of a feather that could fly away with a subtle wind, and which could also be communicated via the use of *fuwafuwa*. AMAI* and FUWAFUWA*, together, create AMAFUWA* to denote the quality of sweet and airy bread.

In contrast, *amafuwa* in (15b) does not refer to the sweetness of food or the texture of bread. Instead, it denotes characteristics of a hairstyle which is very girly and light. So, AMAI is broadened to AMAI** to include something that is not food but is also narrowed, so that it does not include the use of sugar or other sweetener required to make food sweet, for example. Similarly, the expressed concept FUWAFUWA** is narrower than FUWAFUWA, as it is now specific to the context of a hairstyle. Together, it forms the ad hoc concept AMAFUWA that is suitable in this context.

It is interesting that not only are these two constituents of this compositional phrase from different sensory modes (taste and texture), but the composite expression is used to describe items in different modes: one in taste (bread), the other in vision (hairstyle). What is interesting in (15) is the fact that we humans are capable of using such expressions as a tool to communicate our perceptions to each other so that we can share impressions and feelings, which are quite often difficult to put into words. Some onomatopoeia might be more established as words and, of course, a link between sound and meaning is established, as this stems from attempts by humans to recreate their sensory experiences. However, as illustrated in the later examples, humans are even capable of creating new onomatopoeia that go across different sensory domains.

So far, I have shown how notions in relevance-theoretic lexical pragmatics enable us to account for what appears to be the polysemous nature of onomatopoeia. The advantage of this account is that the meaning of onomatopoeia which appears to be polysemous and elusive is, in fact, a result of pragmatic processing, and the intended interpretation by the speaker is on the continuum from literalness to the encoded meaning. Sometimes, it is a case of broadening in the same sensory domain, such as dangling of the doorknob in (11a) and or legs in (11b), or a category extension from the sound (in 13a) to sound and manner (in 13b). Sometimes, it might even include an extension to an abstract level, such as the sound of a rolling object of (13a) to someone's lifestyle in (13c). As these examples show, from a communicative perspective it does not matter whether there is a link between sound and form. It is the hearer who makes a particular connection between sound and meaning in a specific context by modifying the encoded concept. As the relevance-guided comprehension heuristic suggests, this interpretation process is relevance driven: the hearer chooses one meaning over the others because of considerations of relevance.

4 Conclusion: The Semantics and Pragmatics of Onomatopoeia

The central claim of this study is that onomatopoeia is located on the *showing–saying* continuum, where it involves both direct and indirect evidence for the speaker's intention. That is, onomatopoeia is a hybrid of *showing* and *meaning*, providing a cue to ostension via coded and non-coded evidence to a varying degree. This hybrid nature of onomatopoeia allows the representation to activate an impression which it perceptually resembles.

In this chapter, I began by discussing what appears to be the polysemous nature of onomatopoeia. The conceptualist approach often taken in sound-symbolism-based approaches fails to account for how individuals choose one meaning over the other, or fails to explain the subtle differences in meaning and the context dependency.

In contrast, I took a relevance-theoretic approach, an explanatory theory of communication, in analysing the coded aspect of the meaning of onomatopoeia. Drawing on the notions from lexical pragmatics, I argued that what appears to be the polysemous nature of onomatopoeia is a result of lexical modification and ad hoc concept formation in particular. I argued that the lexical concept encoded by (highly lexicalised) onomatopoeia is modified to communicate ad hoc concepts suitable in a specific context, which denote something that is either broader or narrower than the denotation of the lexical concept. This account also provides an explanatory framework for the context dependency of onomatopoeia, as well as the division of labour between semantic and pragmatic aspects of onomatopoeia, often left unexplained in previous studies in Japanese linguistics. Unlike the studies that fail to distinguish between semantic and pragmatic aspects, this study explains the process by which onomatopoeia's lexical concepts are often modified in a specific context. The modification of concepts is guided by the search for optimal relevance. The implication of this is that onomatopoeia are not polysemous after all. That is, the apparent lexical elusiveness of onomatopoeia is a result of pragmatic processing. This, combined with the elusiveness caused by the *showing* aspect of onomatopoeia, as discussed in Chap. 3, makes the meaning of onomatopoeia highly intangible.

So, what is onomatopoeia? Can we say that it forms a word class? The level of lexicalisation would be the key factor in this regard. If certain expressions are located at the *showing* end of the continuum, they would only contribute to relevance by virtue of the perceptual resemblance between their linguistic form and the sensory experience the speaker wishes to communicate by them, whereas other expressions would have an element of an encoded concept, as well as an element of perceptual resemblance. For example, *bang*, as used by English speakers, encodes a concept but also involves activation of certain impressions from its phonology. On the other hand, *tsk*, which could be used mimetically to express someone's irritated state of mind, is an expression without a concept and hence provides direct evidence for the speaker's intention only.[7] This points to the possibility that onomatopoeia is not a class of words. Instead, it is a property of certain linguistic items. This enables us to explain when words behave like an onomatopoeia when they are not. See example (16):

(16) a. Oven kara dashitara *atsuatsu* no uchi ni meshiagare!
Oven from take-out MIM GEN during at eat-HON!
"Once out of the oven, enjoy while piping hot!"

b. Burapi & Jolie sorotte rainichi. *Atsuatsu* kisha kaiken.
Brad Pitt & Jolie together Japan-visit. MIM press conference.
"Brad Pitt and Angeline Jolie visited Japan together. They held a loved-up press conference."
(Eiga.com, 29 November 2005, my italics)[8]

Strictly speaking, the expression *atsuasu* in (16) is not traditionally considered as an onomatopoeia. It is an expression made from the adjective root *atsui*, which is equivalent to *hot* in English. However, the form, which includes duplication of the word stem, is similar to many Japanese onomatopoeia. Indeed, the meaning of *atsuatsu* is highly context dependent, and it also adds a vaguer feeling than the standard adjective *atsui*. For example, in (16a), the use of *atsuatsu* does not simply communicate that the dish is hot. Instead, it brings about a certain impression of a dish that has just come out of an oven, a very steamy dish that might be making a bubbling sound of sizzling sauce. That is, the use of *atsuatsu* does not simply represent the hotness of the dish in terms of the temperature. Instead, it adds excitement to the way the dish is being described.

Similarly, the use of *atsuatsu* in (16b) gives rise to a vague impression of the atmosphere in which the press conference took place. It might be the case that the two actors were holding hands or staring into each other's eyes more than necessary. Alternatively, it might simply be the case that the journalist attending the press conference wanted to emphasise

[7] I am grateful for Kate Scott for suggesting these examples.
[8] https://eiga.com/news/20051129/1/. Accessed 12 May 2019.

the romantic relationship in which the couple was said to be involved. Either way, the use of this expression helps the hearer access such impressions. As an adjective, it would be difficult to say that *atsui* has a *showing* aspect. However, by taking a form that is typical to onomatopoeia, it triggers activation of the prolonged or emphasised state of the event described by the expression. That is, *atsuatsu*, by assuming an onomatopoeic form, now has a *showing* element in addition to the lexical concept encoded by *atsui*. Of course, the lexical concept encoded by *atsui*, or ATSUI, would be modified to mean ATSUI*, according to the context. In the case of (16a), it can be narrowed so that it only involves the temperature of the food, or broadened so that it also denotes what is associated with hot dishes, such as smell and sound. In the case of (16b), the denotation of *atsui*, or ATSUI, is broader than that of the lexical concept, as it has extended its category and means ATSUI** and is now referring to the atmosphere of the couple. Such ad hoc concepts, together with the *showing* aspect, help the hearer to recover highly intangible feelings that the speaker intends to communicate. Similar cases can be found in expressions such as *kodomo-kodomo*, *onnanoko-onnanoko*, or *furifuri*.

As discussed in this chapter, being onomatopoeic does not mean that a word belongs to a word class called onomatopoeia. Instead, onomatopoeia is a property where linguistic forms are used by virtue of their resemblance to the speaker's cognitive experience. Furthermore, being onomatopoeic is also a matter of degree. Some examples are highly lexicalised and have little of an onomatopoeic property, while others have a property of onomatopoeia only and have few coded elements. This might not seem dissimilar to the notion of mimeticity or lexicalisation, as discussed by Tamori and Schorup (1999), Flyxe (2002), or Kadooka (2005). However, the crucial difference is that having an onomatopoeic property does not necessarily convey its iconicity. Nor does it determine the degree of lexicalisation in terms of syntactic features or "semantic" category. Instead, what I have discussed in this chapter as the property of onomatopoeia is firmly cemented within the theoretical architecture of a cognitively grounded theory of communication. That is, the property of onomatopoeia is the human capacity to entertain a perceptual resemblance between their cognitive experience and any tool they can use as a communicative stimulus.

Bibliography

Akita, Kimi. 2009. *A Grammar of Sound-Symbolic Words in Japanese: Theoretical Approaches to Iconic and Lexical Properties of Mimetics*. Kobe: Kobe University.

———. 2013a. Kyoki Tokusei Kara Miru Onomatope No Fureemu Riron. In *Onomatope Kenkyu No Shatei: Chikazuku Oto to Imi*, ed. Kazuko Shinohara and Ryoko Uno, 101–116. Tokyo: Hitsuji Shobo.

———. 2013b. Constraints on the Semantic Extension of Onomatopoeia. *Public Journal of Semiotics* 5 (1): 21–37.

Atoda, Toshiko, and Kazuko Hoshino. 1995. *Giongo/Gitaigo Tukaikata Jiten [Usage Guide to Japanese Onomatopoeias]*. Tokyo: Sotakusha.

Burapi & Jolie Sorotte Rainichi. Atsuatsu Kishakaiken [Brad Pitt and Angelina Jolie Visit Japan Together. Loved-up Press Conference]. 2005. Eiga.Conm. November 29. https://eiga.com/news/20051129/1/.

Carston, Robyn. 1997. Enrichment and Loosening: Complementary Processes in Deriving the Proposition Expressed? In *Pragmatik: Implikaturen und Sprechakte*, ed. Eckard Rolf, vol. 8, 103–127. Linguistische Berichte. Wiesbaden: VS Verlag für Sozialwissenschaften. https://doi.org/10.1007/978-3-663-11116-0_7.

———. 2002. *Thoughts and Utterances: The Pragmatics of Explicit Communication*. Oxford: Blackwell http://discovery.ucl.ac.uk/12106/.

———. 2010. Lexical Pragmatics, Ad Hoc Concepts and Metaphor: A Relevance Theory Perspective. *Italian Journal of Linguistics* 22: 153–180.

Clark, Billy. 2013. *Relevance Theory*. Cambridge: Cambridge University Press. https://doi.org/10.1017/CBO9781139034104.

Diffloth, Gérard. 1972. Notes on Expressive Meaning. *Chicago Linguistic Society* 8: 440–447.

Fillmore, Charles. 1982. Frame Semantics. In *Linguistics in the Morning Calm*, ed. The Linguistics Society of Korea, 111–137. Seoul: Hanshin. https://www.coursehero.com/file/29281284/Fillmore-1982-Frame-Semanticspdf/.

———. 1985. Frames and the Semantics of Understanding. *Quaderni Di Semantica* 6: 222–254.

Flyxe, Martin. 2002. Translation of Japanese Onomatopoeia Into Swedish (with Focus on Lexicalization). *Africa & Asia* 2: 54–73.

Hinton, Leanne, Johanna Nichols, and John Ohala, eds. 1995. *Sound Symbolism*. Cambridge: Cambridge University Press. https://doi.org/10.1017/CBO9780511751806.001.

Inoue, Kazuko. 2013. Onomatope No Tagisei to Sōzōsei [Polysemy and Creativity of Onomatopoeia]. In *OnomatopeKenkyū No Shatei: ChikazukuOto to Imi [Sound Symbolism and Mimetics: Rethinking the Relationship Between*

Sound and Meaning in Language], ed. Kazuko Shinohara and Ryoko Uno, 203–216. Tokyo: Hitsuji Shobo.

Kadooka, Kenichi. 2005. On the Degree of Lexicalization in English Onomatopoeia from a Historical Perspective. *The Ryukoku Journal of Humanities and Sciences* 27 (1): 1–13.

Kindaichi, Haruhiko. 1978. Giongo/Gitaigo Gaisetu [Overview of Mimetics]. In *Giongo/Gitaigo Ziten [A Dictionary of Mimetics]*, ed. Tsuruko Asano, 3–25. Tokyo: Kadokawa Shuppan.

Kita, Sotaro. 1997. Two-Dimensional Semantic Analysis of Japanese Mimetics. *Linguistics* 35 (January): 379–416. https://doi.org/10.1515/ling.1997.35.2.379.

———. 2001. Semantic Schism and Interpretive Integration in Japanese Sentences with a Mimetic: A Reply to Tsujimura. *Linguistics* 39 (January): 419–436. https://doi.org/10.1515/ling.2001.017.

———. 2013. Giongo Gitaigo to Jihatsuteki Miburi. In *Onomatope Kenkyu No Shatei: Chikazuku Oto to Imi*, ed. Kazuko Shinohara and Ryoko Uno, 79–84. Tokyo: Hitsuji Shobo.

Martin, Samuel E. 1975. *A Reference Grammar of Japanese*, Yale Linguistics Series. Vol. 40. New Haven, CT and London: Yale University Press.

Mikami, Kyoko. 2004. Tagi Onomatopoeia No Imi/Yoho No Kijutsu to Shido No Kokoromi: "Gorogoro" "Batabata" o Rei to Shite. *Koide Kinen Nihongo Kyoiku Kenkyukai Ronbunshu* 12: 63–77.

Moriyama, Keiko. 2007. Onomatope no Moto: Kan/Gan [The Foundation of Onomatopoeia: Kan/Gan]. In *Nihongo Onomatope Jiten [Japanese Onomatopoeia Dictionary]*, ed. Masahiro Ono, 57. Tokyo: Shogakukan Publishing.

Nangumo. 2014. Shittori Amafuwa Rennyu-Pan [Moist and Sweet-Airy Milk Bread]. *CookPad*. https://cookpad.com/recipe/2684995.

Ohori, Toshio. 2018. *Ninchi Gengogaku to Goyoron: Gengo Kigo No "Imi" [Cognitive Linguistics and Pragmatics: "Meaning" of Linguistic Symbols]*. Tokyo: Kyorin University.

Ono, Masahiro. 2007. *Nihongo Onomatope Jiten [Japanese Onomatopoeia Dictionary]*. Tokyo: Shogakukan Publishing.

Rohan, Olivia, Ryoko Sasamoto, and Rebecca Jackson. 2018. Argumentation, Relevance Theory and Persuasion: An Analysis of Onomatopoeia in Japanese Publications Using Manga Stylistics. *International Review of Pragmatics* 10 (2): 219–242.

Sasamoto, Ryoko, and Rebecca Jackson. 2016. Onomatopoeia—Showing-Word or Saying-Word? Relevance Theory, Lexis, and the Communication of Impressions. *Lingua* 175: 36–53.

Sasamoto, Ryoko, Minako O'Hagan, and Stephen Doherty. 2017. Telop, Affect, and Media Design: A Multimodal Analysis of Japanese TV Programs. *Television & New Media* 18 (5): 427–440. https://doi.org/10.1177/1527476416677099.

Sperber, Dan, and Deirdre Wilson. 1998. The Mapping Between the Mental and the Public Lexicon. In *Thought and Language*, ed. Peter Carruthers and Jill Boucher, 184–200. Cambridge: Cambridge University Press.

Tamori, Ikuhiro, and Lawrence Clifford Schourup. 1999. *Onomatope: Keitai-to Imi [Onomatopoeia: Form and Meaning]*. Tokyo: Kuroshio Shuppan.

Toratani, Kiyoko. 2005. A Cognitive Approach to Mimetic Aspect in Japanese. *Annual Meeting of the Berkeley Linguistics Society* 31 (1): 335–346. https://doi.org/10.3765/bls.v31i1.856.

———. 2013. Fukushiteki Onomatope No Tokushusei—Tagisei/Jishosei Kara No Kosatsu. In *Onomatope Kenkyu No Shatei: Chikazuku Oto to Imi*, ed. Kazuko Shinohara and Ryoko Uno, 85–99. Tokyo: Hitsuji Shobo.

———. 2018. Semantics and Morphosyntax of Mimetics. In The Cambridge Handbook of Japanese Linguistics, ed. Yoko Hasegawa, 202–221. Cambridge: Cambridge University Press. https://www.cambridge.org/core/books/cambridge-handbook-of-japanese-linguistics/semantics-and-morphosyntax-of-mimetics/25F0EC297C7BB5E257CE270C0CC5CAEB.

Tsujimura, Natsuko. 2001. Revisiting the Two-Dimensional Approach to Mimetics: Reply to Kita (1997). *Linguistics* 39 (2): 409–418. https://doi.org/10.1515/ling.2001.016.

Wharton, Tim. 2009. *Pragmatics and Non-Verbal Communication*. Cambridge: Cambridge University Press. https://doi.org/10.1017/CBO9780511635649.

Wilson, Deirdre. 2003. Relevance and Lexical Pragmatics. *Italian Journal of Linguistics* 15 (January): 273–291.

———. 2004. Relevance and Word Meaning: The Past, Present and Future of Lexical Pragmatics. *Modern Foreign Languages (Quarterly)* 27: 1–13.

———. 2011. Parallels and Differences in the Treatment of Metaphor in Relevance Theory and Cognitive Linguistics. *Intercultural Pragmatics* 8 (2): 177–196. https://doi.org/10.1515/iprg.2011.009.

Wilson, Deirdre, and Robyn Carston. 2006. Metaphor, Relevance and the Emergent Property. *Mind & Language* 21 (3): 404–433.

Wilson, Deirdre, and Robyn Carston. 2007. A Unitary Approach to Lexical Pragmatics: Relevance, Inference and Ad Hoc Concepts. In *Pragmatics*, ed. N. Burton-Roberts, 230–259. Basingstoke: Palgrave.

Wilson, Deirdre, and Dan Sperber. 2002. Relevance Theory. *UCL Working Paper in Linguistics* 14: 249–290.

5

Synaesthesia, Onomatopoeia, and Food Writing

1 Food Writing

Food writing is a genre that provides an interesting insight into the role of onomatopoeia in sharing impressions. Aspects of food writing, of course, involve giving instructions on how to cook certain dishes (cf. Özdamar 2012). However, as Wharton (2010) shows, food writing is much more than a set of instructions—recipes not only provide this, but they "educate and entertain, inform and inspire" (Wharton 2010, 70). Indeed, the food journalist Niamh Mannion says: "[t]he very best food/travel books have the ability to make the reader feel that they have actually visited the locations that are described" (thetaste.ie 2016).[1] Bloom (2008) also claims that "food writers make explicit what native eaters know in their hearts, minds, palates" (2008, 347) and that readers must trust the integrity, authority, and therefore the judgement and taste of the authorial persona—authors must be absolutely reliable. After all, we read food writing because we are "looking for insight, entertainment, relaxation, even more than for information" (Bloom 2008, 354).

[1] http://thetaste.ie/wp/a-taste-of-la-dolce-vita-in-nina-capri-cookbook-review/. Accessed 12 May 2019.

R. Sasamoto, *Onomatopoeia and Relevance*, Palgrave Studies in Sound,
https://doi.org/10.1007/978-3-030-26318-8_5

Despite this consensus on what food writing offers to readers, there are few studies that attempt to explain how food writing goes beyond merely giving a set of instructions and even providing entertainment and inspiration. Instead, studies on food and language have often focused on the linguistic and structural features of recipes, such as the chronological series of steps, agency, the use of the imperative or the first person plural, the omission of the subject (cf. Horsey 1998), speech acts (lack of illocutionary force in imperatives), the layout or structure of recipes (e.g. Cotter 1997), the rhetoric of the recipe (Bloom 2008), the narrative/storytelling structure of food discourse (Norrick 2011; Matwick and Matwick 2014), or language use in food writing and gender (DeVault 1994; Hines 1994, 1999; Counihan 2000; Seto 2003; Hayakawa et al. 2006; Shohet 2007; Takasaki 2012).

Another aspect of food discourse that has attracted scholarly attention is the societal role of food, where a particular focus is given to food and language as a social activity. The French sociologist Bourdieu (1984) argues that food preference often exhibits people's affiliation (or disaffiliation) with communities, as taste in food often depends on "other dimensions of the relationship to the world, to others and to one's own body" (Bourdieu 1984, 193). To Bourdieu (1984, 190), food is "an incorporated principle of classification which governs all forms of incorporation, choosing and modifying everything that the body ingests and digests and assimilates, physiologically and psychologically". One of the most obvious examples of this would be veganism—being a vegan is not just a dietary choice. It is a lifestyle choice and even influences what its followers can wear in terms of clothing and makeup.

Following Bourdieu's approach, scholars such as Szatrowski and colleagues explore the relationship between food and language "as part of a *social activity* through which people construct their worlds by displaying stances (e.g. assessment), identities, shared values, beliefs, etc." (Szatrowski 2014a, 4; original emphasis) in their edited volume (Szatrowski 2014b). For them, food is central to human life, and our identities depend on both food and language. Their research questions, therefore, centre around (1) how humans organise language and bodies around food (i.e. the process of eating, e.g. how ordering at a sushi restaurant is conducted); (2) how we use our language and bodies to taste, identify, and assess food and influence others' preferences, and how these distinctions and discriminations define us as people and construct identities; (3) how food triggers the memory of past experiences and how we talk about our experiences with and tell stories about food and restaurants; and (4) how language and non-verbal behaviour in conversations over and about food socialise us to food practices, affect, taste, gender norms, and identity (Szatrowski 2014a, 4).

Such an approach to food as a "social activity" often focuses on ritual and performance related to food experience, such as serving sushi in restaurants in Japan (Kuroshima 2010, 2014), cross-cultural analysis of commensality (Beeman 2014), socialisation of humans through food, or sociocultural meaning-making via food.[2] One of the dominant methodologies in the analysis of language and food is conversational analysis.[3] Analysing the language use involved in food experience through conversational analysis helps to examine assessments and emotion in food interaction. In addition to the language of food, some conversation analysts (cf. Goodwin and Goodwin 2001; Goodwin et al. 2012) conduct multimodal analyses of food discourse, including speech, bodily expression, gaze, and related artefacts, in order to capture the complex semiotic systems that humans use in interactions. These methodologies allow for analyses of assessment or preference, as well as emotions involved in food discourse. As Goodwin and Goodwin (2001, 254) state, the use of emotion and assessment displays human minds that evaluate their experience.

Unlike scholars who aim to explore the relationship between food and language as a social activity, Wharton (2010) takes a relevance-theoretic approach (Sperber and Wilson 1986/1995) and explains how recipes communicate. Wharton (2010) points out that recipes and food writing do more than just giving a set of instructions for cooking. Food writing educates readers, as well as inspiring them to try something they have

[2] Scholars who focus on social aspects of food and language include Ochs and Shohet 2006; Duranti et al. 2011; Garrett and Baquedano López 2002; Kulick and Schieffelin 2004; Ochs 2002; Ochs et al. 1996; Ochs and Shohet 2006; Schieffelin and Ochs 1986; Clancy 1986; Burdelski 2006; Cook 2006, Cook 2011; DuFon 2006.

[3] Such works include Goodwin 1981, 1986, 2000, 2003, 2007; Karatsu 2004, 2012; Koike 2001, 2009, 2010; Streeck et al. 2011; Szatrowski 2000a, b, 2002a, b, 2003, 2004, 2005a, b, 2008, 2010a, b, c, 2013; Goodwin 1981, 1986; Goodwin and Goodwin 1987, 28–33; Szatrowski 2004, 2013; and others.

never experienced before. As such, recipes must convey what comes with food experience: moods, impressions, or even emotions. According to Wharton (2010, 71), "a recipe is as much about the images and feelings the writer wants to conjure in the mind of the reader as it is about the words it contains, or the way that discourse is set out. We need to *go beyond the words*."

Against this backdrop, the aim of this chapter is to examine the role of onomatopoeia in such a context. How does onomatopoeia contribute to the communication of mood and impressions in food writing? How does onomatopoeia lead the readers to re-enact sensory experience through food? I will apply the relevance-theoretic analysis of onomatopoeia, which is based on the *showing–saying* continuum, as discussed in Chap. 3, to a range of examples of onomatopoeia uses in food discourse. Such examples will include recipe books, food diaries, and product names. However, before discussing this issue, I will first review studies on language uses typically found in food writing.

2 Language in Food Writing

As Szatrowski (2014a) notes, there is a huge body of literature that focuses on the systematic description of the language used in food discourse across languages.[4] Linguistic features often discussed in such studies include chronological series of steps to follow, agency, the use of the imperative, pronoun omission (cf. Horsey 1998), the anti-passive ergative (Wharton 2010), the polysemy of taste adjectives (Muto 2001), verb uses (Newman 2009), synaesthetic and metaphorical expressions (Magee 2009; Kusumi 2005; Majid and Levinson 2011) and the use of onomatopoeia and other terms for food texture in Japanese (Akiyama 2003; Hayakawa 2006; Hayakawa et al. 2005, 2006, 2007; Yoshikawa et al. 1970; Watanabe and Nakamura 2012; Kato et al. 2012a, b; Muto 2003; Hoshino 2015).

Other approaches include the systematic description of expressions related to a specific mode, such as taste or flavour (Seto 2003; Matsuo

[4] See Szatrowski (2014a) for the full list of literature.

2014; Brown 2011; Enfield 2011) or texture (Hayakawa et al. 2004), or specific genres such as menus (Zwicky and Zwicky 1980), recipes (Fisher 1969; Akiyama 2002; Lakoff 2006; Harada 2012; Norrick 2011; Özdamar 2012), television and magazine food advertisements (Koo 2009; Strauss 2005), email discussions about food (Sneijder and te Molder 2005), and gastronomic fiction (Aoyama 2003).

It is particularly interesting to see how the use of onomatopoeia has been studied in relation to food, especially in Japanese linguistics. This is probably due to the nature of onomatopoeia being a bridge between concrete *saying* and more intangible *showing*, as discussed in Chap. 3. In their pioneering work on onomatopoeia in Japanese linguistics, Tamori and Schourup (1999) argue that onomatopoeia gives an impression of "being there" and enables the description of extremely delicate senses. This "being there" or vividness of experience seems to be key to the studies of onomatopoeia. For example, Fujino (2009) argues that the use of onomatopoeia suggests the speaker has an image that cannot be expressed in any direct way.[5] Similarly, in her analysis of onomatopoeia in food writing, Hoshino (2015) argues that onomatopoeia is essentially highly descriptive and suitable for sensory communication.

Hayakawa et al. (1999) take a sound-symbolism approach and argue that onomatopoeia reflects the characteristics of food due to its similarities to sounds/states. Taking a quantitative method and conducting a survey on meanings of onomatopoeia, they first compiled a collection of onomatopoeic expressions as a preliminary study for the systematic description of food properties. The collected onomatopoeia were classified according to their perceived meaning and association with food properties. Hayakawa et al. (1999) then conducted a survey among consumers to confirm whether they agreed with their findings, including the significant difference in onomatopoeia use by gender, gastronomic interest, and frequency of experiencing certain food, but excluding occupational links.

[5] It is not clear what Fujino (2009) means by "expressed in any direct way". Unlike Fujino, our relevance-theoretic analysis considers onomatopoeia as providing both direct and indirect evidence for communication. Perhaps he means "can be expressed in a one-to-one equivalence in words".

Following that study, Hayakawa (2006) continued to characterise onomatopoeia for describing food properties, by focusing on different sensory channels (e.g. appearance, smell, taste, temperature, sound, texture) and an association with specific onomatopoeia. Interestingly, the focus of Hayakawa's (2006) analysis is not just taste. Instead, she emphasises the multisensory nature of the food experience. She argues that food onomatopoeia can be categorised using two criteria, *brittleness* and *elasticity*, as onomatopoeia in food writing often express texture, not just taste, which is related to the sound to which each onomatopoeia is linked via sound symbolism. Hayakawa (2006, 203, my translation) argues:

> As for food expression, it is very difficult to distinguish onomatopoeia and mimetics, as eating involves multiple sensory organs and certain expressions might express multiple sensory expressions in one.

Muto (2003) also discusses the multisensory nature of food experience and shows the usefulness of onomatopoeia in food writing, illustrating links between onomatopoeia and specific foods. She reports that onomatopoeia can address what happens across all five senses and triggers the vicarious experience of food. Similarly, Hoshino (2015) argues that food writing is an attempt to convey texture or taste in a more vivid manner and, as such, requires expressions that differentiate foods that appear to be similar or related to each other. In this study, Hoshino (2015) conducts a qualitative analysis of examples from gourmet magazines, with a focus on novel onomatopoeia and the context that allows for such new expressions. Her findings show that the use of unconventional or novel onomatopoeia allows for a rich description of food experience. Hoshino (2015) also shows how food discourse includes uses of onomatopoeia across all five senses, demonstrating the multisensory nature of food experience. She further shows that onomatopoeia is employed to describe the act of eating, rather than just the food, as the description of the manner of eating also communicates the food experience. Interestingly, she finds that creative onomatopoeia is used less in instructional writing and speculates that this would probably be because the use of non-conventional language could hinder communication. Hoshino (2015) concludes that onomatopoeia is a convenient tool, as by its nature it is highly descriptive, creative, and suitable for sensory communication.

Another aspect of studies on onomatopoeia in relation to food is the user-experience (UX) study. As Kato et al. (2012) say, there has been some work on extracting onomatopoeia from (user-generated) corpus data (cf. Asaga et al. 2008; Uchida et al. 2011). Watanabe and Nakamura (2012, 2349–2350, my emphasis) analysed the use of onomatopoeia on search engines for recipes and state that by "using onomatopoeia words as search keywords, users can find what they want to cook from their *ambiguous* idea". They then compared fitness values between recipe and onomatopoeia based on correlation values between the onomatopoeic word and each keyword in the recipe, including ingredients and actions. The correlation value was calculated by using cooccurrence stats from user-generated recipe corpora. While their findings report interesting evaluation results, which present extremely varied user rankings, they used only a limited range of onomatopoeia, as criticised by Kato et al. (2012). As such, it is difficult to conclude to what extent the correlation between onomatopoeic expressions and recipes can be generalised.

Kato et al. (2012) focus on onomatopoeia in reviews of restaurants and their food. Using a system that extracts onomatopoeic expressions automatically from food review sites, they built an interface where users have access to the list of onomatopoeia specific to the restaurant in which they are interested. They argue that onomatopoeia can describe the food and ambience of the restaurant more directly and delicately and with better realism than merely using general words, which means that onomatopoeia can help with intuitive restaurant searching.

So far, I have reviewed studies that are concerned with onomatopoeia in food writing. It is worth noting that some researchers argue that onomatopoeia can describe the object or experience (food, in case of the food discourse) more directly (Kato et al. 2012) while others (cf. Fujino 2009) claim that onomatopoeia is used when there is no direct expression. However, whatever "direct" means in these studies, a common assumption is that onomatopoeia allows for the communication of a more vivid description. Unfortunately, despite this agreement, these studies focus only on a specific use of onomatopoeia or how there is a possible link between the "sound" of onomatopoeia and food experience. As such, little has been offered to explain *how* and *why* the use of onomatopoeia enables the author to communicate vivid experience, or how it allows for "vicarious" experience, as Muto (2003) puts it. What is similarly interesting, in addition, is the way onomatopoeia in food writing seems to go across different senses, not just taste, but texture, vision, smell, and so on. How does onomatopoeia help us to go across the senses and describe food experience in a vivid manner? In the next section, I will review a recent development in the study of onomatopoeia to examine the links between onomatopoeia and synaesthesia.

3 Synaesthesia, Food, and Onomatopoeia: Re-enacting Food Experience

3.1 Synaesthesia

This section concerns the relationship between onomatopoeia and synaesthesia, with a particular focus on food writing. Simner (2012, 12) characterises synaesthesia as "the pairing of a particular triggering stimulus with a particular resultant experience". That is, synaesthesia is a condition where a certain triggering stimulus results in an extraordinary experience. According to Simner (2012, 1), "a small percentage of the population reports extraordinary sensations of colours, tastes, shapes, etc., triggered by everyday activities such as reading, listening to music, eating, and so on". Synaesthesia takes a number of forms—some people might see certain colours when exposed to music, while others might experience taste upon hearing a particular syllable. Simner (2012) claims that there are at least 61 reported types of manifestation of synaesthesia (Day 2005, 2009, cited in Simner 2012). Historically, synaesthesia has been seen as a cross-sensory or "merging of senses" experience, described in purely sensory-perceptual terms. However, it is now acknowledged that such a definition does not address the majority of synaesthesia that involves higher-order cognitive constructs. Simner, Glover and Mowat (2006a), for example, report that much synaesthesia involves language units such as graphemes and words, which suggests the need for consideration of synaesthesia and linguistic processing.

Synaesthesia, or a "crossing-over" of the senses, has recently started to attract the attention of scholars working on onomatopoeia (Akita 2013). Of course, synaesthesia is not a communicative phenomenon while onomatopoeia is—synaesthesia involves idiosyncratic (and involuntary) links between cognitive domains, while onomatopoeia can be voluntarily exploited by speakers, as discussed in Chap. 3. However, the common characteristic between onomatopoeia and synaesthesia is that both can involve the crossing-over of sensory and linguistic processing. Indeed, as discussed in Chap. 2, there is a huge body of sound-symbolism literature that focuses on the cross-modal nature of onomatopoeia from the perspective of synaesthesia.

As discussed in Sasamoto and Jackson (2016) and again in Chap. 3 of the current study, onomatopoeia involves speakers' exploitation of connections between sensory domains and linguistic processing for communicative effect. This is very similar to the way individuals with synaesthesia create links between senses. This might sound as though onomatopoeia could be taken as a form of synaesthesia, which is not the case at all. It is very important to emphasise that synaesthesia is a condition that is experienced by some part of the population. Onomatopoeia consists of expressions that are used in communication and the use of, or interpretation of, onomatopoeia is not a condition. Synaesthetes cannot control their experience, but the use of onomatopoeia simply depends on the ability and preference of the speaker. Nevertheless, the synaesthetic-like nature of onomatopoeia should not be ignored. The fact that such connections between senses and cognitive constructs can be made suggests that it is not entirely unreasonable that humans could consciously exploit the communicative effect of such connections. In addition, the fact that synaesthesia involves not just sensory and perceptual stimuli but also a higher-order cognitive construct (cf. Simner 2012; Simner, Glover and Mowat 2006a, Simner et al. 2006b to name but a few) suggests that we must go beyond the common perception of onomatopoeia as a word that mimics sounds. That is, onomatopoeia is not only concerned with sound, but also with humans' perceptual and cognitive experience *per se*.

In fact, a number of studies have reported synaesthetic links between phonemes and taste and/or food texture (Day 2005; Simner and Ward 2006; Ward and Simner 2003; Ward et al. 2005; Cytowic 1997; Cytowic and Wood 1982; Kagitani et al. 2014). Gallace et al. (2011, 13) conclude that "our brains are able to extract (perhaps in an automatic fashion) abstract conceptual properties not only from visual and auditory stimuli, but also from food flavour stimuli (despite the fact that from a phylogenetic point of view, taste might be considered a more primitive sense)". It is therefore not surprising that humans take advantage of the existing connection in order to express what they want to communicate, rather than two independent entities having systematic relations, as discussed in Chap. 3. If this is right, then it would suggest that food writers can take advantage of this link to induce a quasi-synaesthetic experience (or enforced synaesthesia), so that readers can feel as though they themselves are having the same food experience as the authors. Indeed, synaesthesia is often linked to creativity (cf. Ward et al. 2008; Mulvenna et al. 2004; Dailey et al. 1997; Domino 1989).

In what follows, I will discuss how exactly humans exploit this link using onomatopoeia in recreating food experience, using a range of examples from food discourse in English and Japanese. But before getting to examples of onomatopoeia, I will examine food discourse from the perspective of relevance theory.

3.2 Food Onomatopoeia and Relevance

Wharton (2010) explains that good recipe and food writing does not simply provide step-by-step instruction. Like any other literary essay genre, it goes beyond so-called procedural discourse. Instead, good recipes and food writing are about emotions and impressions that the author wants to share with their readers. As such, like any other essay genres, "the best food writing evokes impressions, inspires, recalls experience" (Wharton, personal communication) via drawing readers' attention to an array of assumptions that are accessible, or potentially accessible. That is, good food writing communicates impressions and moods, as it involves the author's intention "to bring about in the hearer a marginal increase in the manifestness of a very wide range of assumptions" (Wharton 2010, 72). These assumptions individually may be highly nebulous, but together they allow for the sharing of the mood and atmosphere that surround certain food experiences. The sharing of such impressions enables the reader to recreate such a sensory experience through food.

According to Sperber and Wilson (2015), as discussed in Chap. 3, impressions are a subtype of cognitive experience. Unlike other types of cognitive experience, which can be developed into a fully formed conceptual representation, impressions involve "a change in the manifestness of an array of propositions which all bear on our understanding the same phenomenon, answering the same question, or deciding on the same issue" (Sperber and Wilson 2015, 138). This does not mean that the speaker entertains all propositions that are present in the context. Sperber and Wilson (2015, 138) explain that "an array of propositions have become manifest to you, and although you are not aware of them individually, this overall change in your cognitive environment warrants the inference". That is, this array of propositions as a whole, not individual propositions that form the impression, leads the speaker to a certain conclusion. The question here is how the use of onomatopoeia allows for synaesthesia-like cross-sensory experiences.

Onomatopoeia is often used in creative genres. As I have shown, food writing is a particular genre of procedural discourse that goes beyond providing a set of instructions. The purpose of food writing is not just to inform, but also to share a sensory experience with readers, in order to create the feeling that they have had the same sensory experience. That is, food writing is an attempt to communicate what is essentially intangible, with the aim of provoking readers to re-enact or recreate this experience in their own mind. As experience with food is predominantly sensory, there is little surprise that such experience cannot be fully expressed in words, as writers are required to communicate such non-verbal "feelings" using language. There is thus little wonder that writers choose expressions that allow them to exploit the existing connection between sensory domains and linguistic processing, in order to induce a quasi-synaesthetic experience (or enforced synaesthesia), so that the readers can feel as though they themselves have had the same food experience as the authors.

As I argued in Chap. 3, interpreting onomatopoeia involves a complex process that is driven by considerations of relevance, rather than being a simple case of a matching between a sound and a sense from other domains. As recent studies show (Wharton 2009; Sasamoto and Jackson

2016), onomatopoeia involves both *showing* and *saying* aspects of communication, allowing for communication of impressions (cf. Sperber and Wilson 2015). On the one hand, an established, highly lexicalised element of onomatopoeia (or the wordiness of onomatopoeia) provides indirect, or coded, evidence for communication (and hence contributes to the *saying* end of the spectrum). On the other hand, onomatopoeia's *showing* element provides direct, or non-coded, evidence for communication, where the speaker uses its resemblance to a phenomenon they wish to communicate, bringing about a similar impression that would be perceived at a different sensory domain, and thus inducing enforced synaesthesia among readers. That is, the *showing* aspect of onomatopoeia necessarily involves the exploitation of a perceptual resemblance between two representations, the source phenomenon (i.e. the food experience in this case) and the phonological form of the onomatopoeic expression which the speaker chooses to use to describe the experience. This requires the hearer to determine what properties the food experience and the onomatopoeia share. The commonalities between the food experience and the onomatopoeia will match what is communicated via the *saying* part of onomatopoeia, which may sometimes involve ad hoc concept formation, as discussed in Chap. 4. In addition, what onomatopoeia communicates is often an impression that is so nebulous that it cannot be explained in propositional terms. As a result, the interpretation of onomatopoeia requires a massive amount of inferential reconstruction, and, as such, interpreting onomatopoeia is extremely context dependent.

Onomatopoeia is a case of humans' exploitation of connections over different sensory domains, allowing for the communication of extremely intangible and vague impressions, in addition to propositional content provided by the *saying* aspect. Onomatopoeia, therefore, provides quick and dirty access to (extremely weak) non-propositional effects (i.e. impressions) that are based on sensory experience, which in turn allows the readers to feel as though they have undergone the food experience themselves, via enforced synaesthesia.

Note that the term *enforced synaesthesia* does not mean that the use of onomatopoeia makes the reader taste in the real sense of synaesthesia. After all, the purpose of this study is to account for the role of onomatopoeia in communication, and the study of onomatopoeia in terms of synaesthesia does not do this, hence it limits our understanding of communication. Furthermore, analysing onomatopoeia in terms of synaesthesia does not account for the crossover of more than two senses. Food experience often involves more than two sensory modes; in fact, it is highly multimodal and frequently involves all five senses: vision, audition, gustation, olfaction and somatosensation (or tactile sense), as Hoshino (2015) points out. However, the study of onomatopoeia in terms of synaesthesia, in turn, provides an insight into how humans exploit connections between language and other senses.

In what follows, examples from a range of food writing in both English and Japanese will be examined, in order to show that humans can take advantage of linguistic processing to share a sensory experience; how onomatopoeia contributes to the communication of impressions in this particular genre; and how the use of onomatopoeia enables the reader to "experience" food to the point that they would be tempted to re-enact it (e.g. to cook the dish, or to visit the restaurant).

3.3 Onomatopoeia in Food Writing

So far, I have argued that food writing is an attempt to verbalise what is essentially too intangible for readers to re-enact or recreate in their own mind. The purpose of food writing is, therefore, to lead the reader to act on what they have read, and onomatopoeia helps readers to experience senses associated with food, including taste/texture/smell, via direct evidence provided through different sensory channels as well as via indirect (coded) evidence for the interpretation given through the *saying* part of the expression.

Nigella Lawson is a famous chef who is well known for her evocative, sensual language and therefore it is not surprising that onomatopoeia is prevalent in her writing:

(2) There is something so welcoming about a big bowlful, the rich, smooth, eggy cream waiting to ***ooze*** out on the spoon that breaks through the tortoiseshell disc on top.
(Nigella Lawson, "Crème Brulée," 2001, my italics)[6]

[6] https://www.nigella.com/recipes/creme-brulee. Accessed 12 May 2019.

Example (1) includes the use of the onomatopoeic expression *ooze out*. This shares certain characteristics with the chocolate used in this recipe. The description given by the chef tells you how thick and creamy the chocolate is. The definition of *ooze* given by the *OED* is to "emit slowly". However, the use of *emit slowly* instead of *ooze out* in this context would not give rise to the same non-propositional effects, although both expressions might convey something similar to "come out slowly". In addition, the phonological form of *ooze* (the long vowel in particular) resembles, or shares a certain quality with, the manner in which we experience the chocolate slowly coming out of the shell. The low pitch of the vowel also provides a cue for a thick, creamy sauce. This aspect of *ooze* provides quick and dirty access to the impression of how cream emerges when the hard surface of melted sugar is broken, which would typically be experienced through vision and/or somatosensation. That is, the onomatopoeic expression provides direct evidence of this particular experience via the *showing* aspect of the phrase using the sound, giving rise to the non-propositional effects that form an impression. In doing so, the onomatopoeic expression enables the reader to have the quasi-experience of this particular food via enforced synaesthesia between vision and/or somatosensation and audition.

Example (2) is taken from a recipe for a crisp aubergine with chilli sauce and illustrates how different onomatopoeic expressions can be used to describe the food preparation process, as well as the state of the food:

(2) Cook for 5 minutes or until the outer batter is ***crisp***, the inner flesh soft as a marshmallow. Drain each piece briefly on kitchen paper then ***trickle*** with the hot chilli sauce and ***sprinkle*** over some of the sesame seeds, a little coriander for those who like it, and perhaps small, ripe chilli, sliced as thin as tissue paper.

(Nigel Slater, "Nigel Slater's Crisp Aubergines with Chilli Sauce Recipe", 2018, my italics)[7]

Here, the expressions *crisp*, *trickle*, and *sprinkle* are used to describe different aspects of the cooking. *Crisp*, which is defined in the *OED* as "[b]rittle or 'short' while somewhat hard or firm in structure (usually as a

[7] https://www.theguardian.com/lifeandstyle/2018/feb/20/nigel-slater-crisp-aubergines-with-chilli-sauce-recipe. Accessed 12 May 2019.

good quality); said esp. of hard things which have little cohesion and are easily crushed by the teeth", describes how the aubergine should be cooked and how it should come out after being deep-fried for five minutes. Here, in addition to describing the state of the aubergine after deep-frying, the use of *crisp* evokes certain impressions, or gives rise to extremely nebulous effects, about how aubergine should be hard or firm in structure. Such an impression leads to a re-enactment (quasi-experience) of certain sensations in the reader's mouth, as though the aubergine of this particular texture were to go into their mouth. *Crisp* provides such direct evidence of this particular experience via its *showing* element, in addition to the *saying* element, as described earlier.

Unlike *crisp*, *trickle* and *sprinkle* are used to instruct the reader how to physically perform certain actions within the cooking process. In addition to *emit or* "give force in successive drops, or pour drop by drop", as defined in the *OED*, *trickle* enables the author to communicate the manner in which chilli sauce is being emitted in successive drops. In this case, the use of the expression will lead the reader to experience how the chilli sauce must be added on top of the aubergine, in a very careful, tiny stream of drops, via the phonological form that resembles the experience. The use of *sprinkle* also instructs the reader exactly how to "scatter in small amounts", as defined in the *OED*. Not only does this expression have the *saying* element that instructs the reader to scatter the sesame seeds in a small amount, but it also *shows* the lightness of scattering, as it provides the cue for the experience via its phonological form that resembles the experience. Each cookery action would normally be experienced through vision or somatosensation. However, even when there is no means of using vision or touch, the use of onomatopoeia will enable the reader to have this particular experience, as it provides direct evidence via its resemblance to the experience and, as a result, triggers enforced synaesthesia.

Next is the onomatopoeic expression *slurp*, which is used to describe the manner of eating, rather than cooking:

(3) I love the Japanese way of eating cold noodles: I just lift a bowl to my face, fork furiously and ***slurp***.

(Nigella Lawson, "Soba Noodles with Sesame Seeds", 2012, my italics)[8]

[8] https://www.nigella.com/recipes/soba-noodles-with-sesame-seeds. Accessed 12 May 2019.

(4) The principal pleasure of cacio e pepe is in the ***slurping*** up of slippery strands of saucy pasta.
(Felicity Cloake, "How to Make a Perfect Cacio e Pepe", 2016, my italics)[9]

In both (3) and (4), the onomatopoeia *slurp* provides easy access for the reader to a wide range of non-propositional effects associated with this particular way of eating noodles/pasta. This manner of eating involves some noise (sound) as well as the somatosensation of the noodles going through the mouth to the throat. By making such non-propositional effects more accessible to readers via its phonological form, the onomatopoeia enables them to feel as though they are actually experiencing the food via enforced synaesthesia.

The expression *slurp* is particularly interesting, as it clearly illustrates the difference between onomatopoeic expressions and their paraphrase in conceptual terms. *Slurp* is defined in the *OED* as "to drink or eat greedily or noisily". However, this somewhat distasteful definition does not seem appropriate in some contexts, as the word can also be used to communicate a favourable food experience. As illustrated in examples such as (3) and (4), the onomatopoeia *slurp* can be used to communicate how eating noodles/pasta in a noisy manner can create pleasure. Replacing *slurp* in (3) and (4) with *to eat greedily or noisily* as in (3b) and (4b) will not be interpreted in the same way, and, indeed, such a paraphrase would make both utterances particularly awkward:

(3a) I love the Japanese way of eating cold noodles: I just lift a bowl to my face, fork furiously and ***slurp***.
(Nigella Lawson, "Soba Noodles with Sesame Seeds", 2012, my italics)
(3b) ?? I love the Japanese way of eating cold noodles: I just lift a bowl to my face, fork furiously and ***eat greedily or noisily***.
(4a) The principal pleasure of cacio e pepe is in the ***slurping*** up of slippery strands of saucy pasta.
(Felicity Cloake, "How to Make a Perfect Cacio e Pepe", 2016, my italics)
(4b) ?? The principal pleasure of cacio e pepe is in the ***eating greedily or noisily*** of slippery strands of saucy pasta.

[9] https://www.theguardian.com/lifeandstyle/wordofmouth/2016/nov/03/how-to-make-the-perfect-cacio-e-pepe. Accessed 12 May 2019.

As shown in the paraphrase of *slurp* in both examples, replacing the word with its conceptual counterpart does not result in a similarly pleasant description of eating noodles. In both cases, the somewhat pleasant feeling of satisfaction is taken away when *slurp* is replaced with *eating greedily or noisily*. The question is, why can we not simply replace the onomatopoeic expression with a dictionary-defined alternative phrase? The answer lies in the way onomatopoeia involves both *showing* and *saying* elements. Earlier, I argued that onomatopoeia's *showing* element provides non-coded, direct evidence for communication that brings about a range of extremely weak non-propositional effects that form impressions. The paraphrase using a conceptual term (in this case, *eat greedily or noisily*) might provide an alternative for the *saying* aspect of *slurp*. However, it is too restrictive, in that such a paraphrase does not allow for the communication of "a wide array" of propositions. Indeed, as discussed earlier, the speaker does not entertain all propositions that form the impression (cf. Sperber and Wilson 2015). So, in certain contexts, *slurp* might be used to describe a manner of eating that is distasteful negatively and, in that case, the more negative range of propositions would be made manifest. In other cases, such as (3) and (4), a more favourable subset of propositions would be made manifest and entertained. This shows how the use of certain conceptual terms can be too restrictive in a certain context that requires the evocation of moods, such as food discourse, and alternative, less tangible expressions such as onomatopoeia would be more suitable, since these allow for the communication of impressions.

Both the expressions I have analysed so far have some direct link to the food they are used to describe. *Ooze out* denotes the way cream comes out of its shell, while *slurp* denotes an actual manner of eating. In both cases, onomatopoeia, using its phonological form, represents aspects of how the author perceives the experience in one (or more) sense(s). Next, I will analyse a case where the author uses an onomatopoeic expression that does not normally have a direct link to the food experience:

(5) A super-reliable recipe for minced pork stuffing—gives your roast dinner some ***oomph***.

(Jamie Oliver "Stuffing", 2013, my italics)[10]

[10] https://www.jamieoliver.com/recipes/pork-recipes/stuffing/. Accessed 12 May 2019.

This is an example from TV chef Jamie Oliver, who is famous for the use of casual and descriptive language. In this example, in order to describe what the roast dinner is like, he uses the onomatopoeia *oomph*. The definition of *oomph*, according to the *OED*, is "[t]he quality of being exciting, energetic, or sexually attractive; energy, vigour". Under normal circumstances, the object of the verb phrase in the utterance *gives your roast dinner some X* would be something that is related to flavour or smell. However, in this case, what minced pork stuffing gives is beyond a simple flavour or smell. Instead, if we follow the *OED* definition, it gives *the quality of energy or excitement*. What does the author intend to communicate here?

The relevance-theoretic analysis presented in this study predicts that *oomph* here enables readers to access extremely weak, non-propositional effects that form impressions. Such effects involve an array of a wide range of propositions which the speaker might not entertain fully, and the *showing* aspect of the expression makes such an impression manifest to the readers. So, what does *oomph* make manifest? Rather than a certain smell or taste that is often associated with food experience, this onomatopoeia makes manifest impressions that are less specific in terms of particular senses. In contrast to *slurp* in (2) and (3), which gives rise to non-propositional effects associated with the experience of the actual eating process, *oomph* in (4) gives rise to non-propositional effects that go beyond the actual (i.e. physical) process of eating and apply to the dish as a whole. It encompasses a strength in flavour, long-lasting taste, the taste of the stuffing itself, and how it affects the roast meat, combined with the smell, and of course the quality of the dish as a whole, combining all senses available.

Next, example (6) shows how an onomatopoeic expression *glug* is used to describe the action of to "pour or drink (liquid) with a hollow gurgling sound" (*OED*). Here, instead of describing someone gurgling, it is used to describe the amount of liqueur used in baking:

(6) The cakes were finished with a frosting made from mascarpone and enriched with an egg yolk, a little icing sugar and a generous ***glug*** of Amaretto—an Italian almond flavoured liqueur.

(Niamh Mannion, "A Taste of La Dolce Vita in NINA Capri—Cookbook Review", my italics)[11]

[11] http://thetaste.ie/wp/a-taste-of-la-dolce-vita-in-nina-capri-cookbook-review/. Accessed 12 May 2019.

This is an extract from a review of a cookbook, *NINA Capri: Recipes from Italy's Amalfi Coast*, by Nina Parker, a London-based chef and food writer. The author, according to the reviewer of the book on this website, takes inspiration from her childhood in the South of France. The use of *glug* in this example provides a perfect example of onomatopoeia in food writing and it is reasonable to say that it is not only used to replicate the sound of pouring the liqueur from the bottle. Instead, the use of *glug* probably gives rises to an impression of an unsophisticated (not in a negative but in a good way), not oversensitive, or simply relaxed atmosphere. This reflects what the reviewer says about the chef taking inspiration from her childhood. The use of *glug* in this example probably gives rise to non-propositional effects such as the memory of uncalculated, homely cooking, nostalgia from childhood, a relaxed, unassuming atmosphere, and so on. Onomatopoeia in food writing not only provides a clue about what food is like, but also allows the reader to feel the author's experience as if it were their own.

The fact that onomatopoeia is located on the continuum of *showing* to *saying* enables us to explain the difference between closely related onomatopoeic expressions. Take the example of expressions that are linked with gustatory satisfaction or pleasure. Such expressions would include *nom nom*, *mmm*, *ymmm*, *yum*, *yummy*.[12] Interestingly, while all these expressions can be used to demonstrate gustatory pleasure or satisfaction, not all are listed in the dictionary, which suggests that they exhibit varying degrees of lexicalisation. For instance, *yummy* is highly lexicalised and defined in the *OED* as "an exclamation of pleasurable anticipation, with the implication of sensual or gustatory satisfaction". It is even part of the newly coined expression *yummy mummy*, further suggesting that it is highly lexicalised and hence has a stronger *saying* element. Other expressions such as *nom nom* or *ymmm* do not feature in the *OED*. *Mmm* (or its variation *mm* or *mmmmm*) features in some dictionaries (e.g. the *OED*) but not in others, suggesting that it is not as established a lexical item as *yummy* but, at the same time, not entirely novel. Being listed in dictionaries does not necessarily equate to an expression's status as a word, but it would certainly indicate how each form is recognised as lexis. The question is whether or not such differences in the degree of lexicalisation

[12] I am grateful for Kate Scott for suggesting this fascinating range of expressions.

would make any difference to communication, as all of them are used to indicate the speaker's satisfaction with food, after all. If so, shouldn't they be more or less the same? Intuitively, *nom nom* or *ymmm* seems to reflect the speaker's reaction to the food in question, when the speaker is deeply engaged in the experience or making a somewhat emotionally driven judgement, while *yummy* or *yum* communicate somewhat distant judgements made objectively by the speaker. That is, expressions such as *nom nom* or *ymmm* appear to reflect the speaker's natural response to the food, while *yum* or *yummy* convey a somewhat more analytic impression. As discussed in Chap. 3, onomatopoeia is located on the *showing–saying* continuum, providing both direct and indirect (coded) evidence for communication. Some expressions might have a stronger *showing* aspect, while others are located more towards the *saying* end of the spectrum. This would predict that *nom nom* or *ymmm* is located more towards the *showing* end of the continuum and is therefore less lexically established than *yummy* or *yum*, while *yummy* or *yum* is located nearer to the *saying* end of the continuum and is therefore established as a lexeme. This would explain why the latter would give a somewhat objective impression.

3.4 Onomatopoeia and Product Naming

As discussed throughout this book, the nature of onomatopoeia means that its use allows for the sharing of impressions via the resemblance between the phonological form and the cognitive experience the speaker wishes to communicate. Together with its verbal element, which provides indirect, or coded, evidence for communication, onomatopoeia helps to communicate nebulous impressions that are otherwise very difficult to convey. This very nature of sharing impressions is therefore extremely convenient for product naming. It is therefore unsurprising that the use of onomatopoeia is prevalent across advertising slogans and product naming (Abdul Razak 2012; Gherasim 2014). In this section, I will examine a number of product names and advertising slogans to show how onomatopoeia is used to communicate the quality of each product.

Let me first take the examples of Rice Krispies and Coco Pops, two breakfast cereals made by Kellogg's which have been on the market since the early twentieth century. Both are made of rice, puffed and made crispy outside. The focus on the onomatopoeic sound in the advert for Rice Krispies is well known, including the advertising slogan *snap, crackle and pop*, which seemingly is based on the noise the cereal makes when mixed with milk, as well as the TV adverts that feature children listening to the sound when the mother pours milk into a bowl of Rice Krispies. The product name, Rice Krispies, also involves onomatopoeic aspects in *krispie*, although the spelling of the name *Krispies* is changed from the more conventional /c/ to /k/, which again happens often in product naming (Lischer 2015). So, when presented with this product and its advertisement, consumers can access a wide range of impressions, including those from its name—that is, the manner in which the cereal is so very crispy—and the changing texture, which is communicated through the use of three onomatopoeic expressions: *snap*, *crackle*, and *pop*. Altogether, these expressions enable consumers to access a range of non-propositional effects that amount only to impressions of a particular sensation in the mouth with the texture this particular breakfast cereal will give. This may involve some sort of cracking (or crackling) noise, or at least one would not expect something smooth or stodgy.

Coco Pops, like Rice Krispies, have an onomatopoeic element in their name. The name Coco Pops, and the use of *pop* in particular, gives rise to an impression of a crispy, cocoa-flavoured breakfast cereal which allows you to "experience" the particular sensation and texture of eating this cereal. If the name were Cocoa Cereal instead, it would be more difficult to imagine what the cereal would be like.

Whether Rice Krispies or Coco Pops, the use of onomatopoeic expressions provides consumers with access to what the breakfast cereal is like via direct evidence, which is based on the resemblance between the phonological aspect of the word *pop* and the experience of eating the cereal. That is, the consumer would be able to imagine the quality (or texture) of this particular product via just the product's name and the slogans used in the advertising campaign.

Another brand name with the expression Krispie is Krispie Kreme, a US-born doughnut chain. The name Krispie Kreme, especially the use of *krispie* (regardless of misspelling), evokes a certain impression of doughnuts that are crispy to bite into. When presented with the name of the brand, consumers can access assumptions about doughnuts with a glaze that gives a crispy bite; how the first bite might feel; and how the glaze might crackle across the surface, as the non-coded aspect of the onomatopoeia *crispy* provides direct evidence for this, via its resemblance to the texture.

It is not just the word *crispy* that is often used in food names. *Munch* is another word that frequently appears, including in *Monster Munch*, a corn snack made by the British company Walkers, as well as *Mighty Munch*, manufactured by the Irish company Tayto. Both snacks have an image of a monster on the package. The use of the onomatopoeia *munch* helps communicate what it is like to consume the snack. The phonological form of the word *munch* perceptually resembles a particular manner of eating, which leads to the recovery of certain impressions. That is, the use of the word *munch* provides consumers with access to an array of extremely weak assumptions about a certain way of consuming food that goes beyond the simple act of eating. It would involve assumptions about not being able to stop, eating with enthusiasm with a loud noise, and so on, just like monsters would do. That is, the use of onomatopoeia helps to communicate how exciting it would be to eat the snack.

Sometimes, the use of onomatopoeia in product naming is not limited to the name of the product itself but applies to a whole brand. Let us take the example of *Sizzler*, an American-style restaurant chain with restaurants throughout the world, including in Japan, Australia, Thailand, and Puerto Rico. As the name indicates, the restaurant is famous for its steak and salad bar. The name Sizzler obviously comes from the noise of cooking steaks. The use of onomatopoeia helps consumers to access impressions about the dishes, that they are sizzling, and what it would be like to be in a restaurant buzzing with customers eating such steaks. Again, this is made possible as the use of onomatopoeia allows for the sharing of impressions via the perceptual resemblance between the phonological form of the word *sizzler* and the cognitive experience of steak being cooked on a hot pan.

Of course, the brands analysed here are all well established and it is unlikely that a consumer would need to be introduced to them as a new product. However, the use of onomatopoeic expressions in their advertising campaigns and their names would certainly give rise to impressions of these products that would appeal to consumers' pleasant experiences of eating the cereals or snacks, which might lead to a further purchase. That is, whether or not a customer has previous experience of each product (or restaurant), the use of onomatopoeia in its name enables the customer to experience for the first time, or to be reminded of their experience of, the product via its perceptual resemblance to the perception of the product.

4 Conclusion

The aim of this chapter was to analyse the role of onomatopoeia in food discourse and to explain how the use of onomatopoeia helps the reader to experience enforced synaesthesia via food writing. I have shown how the characteristics of onomatopoeia with both *showing* and *saying* aspects allow the reader to access both direct and indirect (coded) evidence of what the author intends to communicate. The onomatopoeic expressions in the examples given clearly illustrate how food writing must "go beyond the words" (Wharton 2010, 71) and take readers right into the midst of the food experience (cf. Mannion 2016). As I have argued, paraphrasing onomatopoeic expressions with their counterparts, which consist solely of the *saying* aspect (or are solely verbal), is too restrictive to fully express the impressions and feelings of food experience. This is not particularly surprising, as food experience often goes across different senses and beyond, and, as such, authors would presumably need to appeal to more emotive and less concrete terms. Onomatopoeia is a perfect tool for authors, in that it allows for the enforced synaesthesia of the full multisensory experience of food. Onomatopoeia provides quick and dirty access to (extremely weak) non-propositional effects (i.e. impressions) that are based on sensory experience, which in turn allow readers to feel as though they have undergone the food experience themselves, via enforced synaesthesia.

Bibliography

Abdul Razak, Siti Hajjar. 2012. Japanese Onomatopoeia: Cosmetics and Skincare Products in Fashion Magazines. *Center for Comparative Japanese Studies Annual Bulletin* 8: 123–128.

Akita, Kimi. 2013. Onomatope/Otoshocho No Kenkyu-Shi. In *Onomatope Kenkyu No Shatei: Chikazuku Oto to Imi*, ed. Kazuko Shinohara and Ryoko Uno, 333–364. Tokyo: Hitsuji Shobo.

Akiyama, Madoka. 2002. "Syoku" o Meguru Hyoogen- Resipi No Bunsyoo/Buntai Ni Tuite [Expressions Related to "Food": On the Discourse/Style of Recipes]. *Kokugakuin Daigaku Kokugo Kenkyuukai "Kokugo Kenkyuu" [Kokugakuin University Japanese Research Society "Inquiries into the Japanese Language"]* 65: 34–46.

———. 2003. Syokkan o Arawasu Onomatope Ni Tuite [On the Onomatopoeia of the Texture]. *Kokugakuin Daigaku Kokugo Kenkyuukai "Kokugo Kenkyuu" [Kokugakuin University Japanese Research Society "Inquiries into the Japanese Language"]* 66: 12–27.

Aoyama, Tomoko. 2003. The Cooking Man in Modern Japanese Literature. In *Asian Masculinities: the Meaning and Practice of Manhood in China and Japan*, ed. K. Louie and M. Low, 155–176. London: Routledge Curzon.

Asaga, Chisato, Yusuf Mukarramah, and Chiemi Watanabe. 2008. ONOMATOPEDIA: Onomatopoeia Online Example Dictionary System Extracted from Data on the Web. In *Proceedings of the 10th Asia-Pacific Web Conference on Progress in WWW Research and Development*, 601–612. APWeb'08. Berlin and Heidelberg: Springer-Verlag. http://dl.acm.org/citation.cfm?id=1791734.1791805.

Beeman, William O. 2014. Negotiating a Passage to the Meal in Four Cultures. In *Pragmatics & Beyond New Series*, ed. Polly E. Szatrowski, vol. 238, 31–52. Amsterdam: John Benjamins Publishing Company. https://doi.org/10.1075/pbns.238.02bee.

Bloom, Lynn Z. 2008. Consuming Prose: The Delectable Rhetoric of Food Writing. *College English* 70 (4): 346–362.

Bourdieu, Pierre. 1984. *Distinction—Pierre Bourdieu | Harvard University Press*. Cambridge, MA: Harvard University Press http://www.hup.harvard.edu/catalog.php?isbn=9780674212770.

Brown, Penelope. 2011. Color Me Bitter: Crossmodal Compounding in Tzeltal Perception Words. *The Senses and Society* 6 (1): 106–116. https://doi.org/10.2752/174589311X12893982233957.

Burdenski, Matthew James. 2006. *Language Socialization of Two-Year Old Children in Kansai, Japan: The Family and Beyond*. Ann Arbor, MI: UCLA.

Clancy, Patricia M. 1986. The Acquisition of Communicative Style in Japanese. In *Language Socialization Across Cultures*, ed. Bambi B. Schieffelin and Elinor Ochs, 213–250. New York: Cambridge University Press.

Cloake, Felicity. 2016. How to Make the Perfect Cacio e Pepe. *The Guardian*, November 3, sec. Food. https://www.theguardian.com/lifeandstyle/wordofmouth/2016/nov/03/how-to-make-the-perfect-cacio-e-pepe.

Cook, Haruko Minegishi. 2006. Joint Construction of Folk Beliefs by JFL Learners and Japanese Host Families. In *Language Learners in Study Abroad Contexts*, ed. Margaret A. DuFon and Eton E. Churchill, 120–150. Bristol: Multilingual Matters. https://doi.org/10.21832/9781853598531-009.

———. 2011. Language Socialization and Stance-Taking Practices. In *The Handbook of Language Socialization*, ed. Alessandro Duranti, Elinor Ochs, and Bambi B. Schieffelin, 296–321. Malden, MA: Blackwell. https://doi.org/10.1002/9781444342901.ch13.

Cotter, Colleen. 1997. Claiming a Piece of the Pie: How the Language of Recipes Defines Community. In *Recipes for Reading: Community Cookbooks, Stories, Histories*, ed. Anne Bower, 51–71. Amherst, MA: University of Amherst Press.

Counihan, Carole M. 2000. Fantasy Food: Gender and Food Symbolism in Preschool Children's Madeup Stories. In *The Anthropology of Food and Body: Gender, Meaning and Power*, ed. Carole M. Counihan, 129–155. Longon: Routledge.

Cytowic, Richard E. 1997. Synesthesia: Phenomenology and Neuropsychology—A Review of Current Knowledge. In *Synaesthesia: Classic and Contemporary Readings*, ed. John E. Harrison and Simon Baron-Cohen, 17–42. Oxford: Blackwell.

Cytowic, Richard E., and Frank B. Wood. 1982. Synesthesia: II. Psychophysical Relations in the Synesthesia of Geometrically Shaped Taste and Colored Hearing. *Brain and Cognition* 1 (1): 36–49. https://doi.org/10.1016/0278-2626(82)90005-7.

Dailey, Audrey, Colin Martindale, and Jonathan Borkum. 1997. Creativity, Synesthesia and Physiognomic Perception. *Creativity Research Journal* 10 (1): 1–8. https://doi.org/10.1207/s15326934crj1001_1.

Day, Sean. 2005. Some Demographic and Socio-Cultural Aspects of Synesthesia. In *Synesthesia: Perspectives From Cognitive Neuroscience*, ed. Lynn C. Robertson and Noam Sagiv, 11–33. Oxford: Oxford University Press.

———. 2009. Types-of-Syn. http://www.daysyn.com/Types-of-Syn.html.

DeVault, Marjorie L. 1994. *Feeding the Family*. Chicago: University of Chicago Press. https://www.press.uchicago.edu/ucp/books/book/chicago/F/bo3684531.html.

Domino, George. 1989. Synesthesia and Creativity in Fine Arts Students: An Empirical Look. *Creativity Research Journal* 2 (1–2): 17–29. https://doi.org/10.1080/10400418909534297.

DuFon, Margaret A. 2006. The Socialization of Taste during Study Abroad in Indonesia. In *Language Learners in Study Abroad Contexts*, ed. Margaret A. DuFon and Eton E. Churchill, 120–150. Clevedon: Multilingual Matters. https://doi.org/10.21832/9781853598531-008.

Duranti, Alessandro, Elinor Ochs, and Bambi B. Schieffelin, eds. 2011. *The Handbook of Language Socialization*. Malden, MA: Blackwell. https://doi.org/10.1002/9781444342901.ch13.

Enfield, Nicholas J. 2011. Taste in Two Tongues: A Southeast Asian Study of Semantic Convergence. *The Senses and Society* 6 (1): 30–37. https://doi.org/10.2752/174589311X12893982233632.

Fisher, Mary Frances Kennedy. 1969. The Anatomy of a Recipe. In *With Bold Knife and Fork*, 1–14. New York: Counterpoint.

Fujino, Yoshitaka. 2009. Tensai Asuriito No Oto [The Sound of Genius Athletes]. *Watashi No Onomatope Sutairu [My Onomatopoeia Style]*. http://happy-vocalism.cocolog-nifty.com/blog/cat41995244/.

Gallace, Alberto, Erica Boschin, and Charles Spence. 2011. On the Taste of "Bouba" and "Kiki": An Exploration of Word–Food Associations in Neurologically Normal Participants. *Cognitive Neuroscience* 2 (1): 34–46. https://doi.org/10.1080/17588928.2010.516820.

Garrett, Paul B., and Patricia Baquedano-López. 2002. Language Socialization: Reproduction and Continuity, Transformation and Change. *Annual Review of Anthropology* 31 (1): 339–361. https://doi.org/10.1146/annurev.anthro.31.040402.085352.

Gherasim, Toader. 2014. The Concept of a Brand. *Economy Transdisciplinarity Cognition* 17: 8–12.

Goodwin, Charles. 1981. *Conversational Organization: Interaction between Speakers and Hearers*. New York: Academic Press.

———. 1986. Between and within: Alternative Sequential Treatments of Continuers and Assessments. *Human Studies* 9 (2): 205–217. https://doi.org/10.1007/BF00148127.

———. 2000. Action and Embodiment within Situated Human Interaction. *Journal of Pragmatics* 32: 1489–1522.

———. 2003. Pointing as Situated Practice. In *Pointing: Where Language, Culture, and Cognition Meet*, ed. Sotaro Kita, 217–241. Mahwah, NJ: Lawrence Erlbaum Associates.

———. 2007. Environmentally Coupled Gestures. In *Gesture and the Dynamic Dimension of Language*, ed. Susan D. Duncan, Justine Cassell, and Elena T. Levy, 195–212. Amsterdam: John Benjamins Publishing. https://benjamins.com/catalog/gs.1.18goo.

Goodwin, Charles, and Marjorie Harness Goodwin. 1987. Concurrent Operations on Talk: Notes on the Interactive Organization of Assessments. *IPRA Papers in Pragmatics* 1 (1): 1–54.

Goodwin, Marjorie Harness, and Charles Goodwin. 2001. Emotion within Situated Activity. In *Linguistic Anthropology: A Reader*, ed. Alessandro Duranti, 239–257. Oxford: Blackwell.

Goodwin, Marjorie, Asta Cekaite, and Charles Goodwin. 2012. Emotion as Stance. In *Emotion in Interaction*, ed. Anssi Perakyla and Marja-Leena Sorjonen, 16–40. Oxford: Oxford University Press. https://doi.org/10.1093/acprof:oso/9780199730735.003.0002.

Harada, Aya. 2012. Resipi No Buntai Ni Kansuru Kenkyuu- Zassi "Huzin No Tomo" No Ryoori Kizi o Taisyoo Ni- [Research on the Style Used in Recipes: Recipe Articles in the Magazine Huzin No Tomo "Women's Friend"]. *Otyanomizu Zyosi Daigaku Kokugo Kokubun Gakkai "Kokubun" [Ochanomizu University Kokugo Kokubun Society "National Literature"]* 17: 51–61.

Hayakawa, Fumiyo. 2006. *Syokkan No Onomatope [Onomatopoeia for Food Texture]*. Tokyo: Sanseido.

Hayakawa, Fumiyo, Yufuko Iwamasa, Keiko Hatae, and Atsuko Shimada. 1999. Collection of Onomatopoeic Terms for Describing Food Properties. *Nihon Kasei Gakkaish [Journal of Home Economics Japan]* 50 (5): 481–491.

Hayakawa, Fumiyo, Shun-Sheng Chen, Xi-Chang Wang, Zai-Qui Li, Masayoshi Saito, Yasumasa Baba, and Masahito Yokoyama. 2004. Collection and Classification of Chinese Texture Terms: Studies on Chinese Texture Terms. *Nippon Syokuhin Kagaku Koogaku Kaisi [Food Science and Technology Research]* 51 (3): 131–141.

Hayakawa, Fumiyo, Kana Ioku, Sayuri Akuzawa, Masayoshi Saito, Katsuyoshi Nishinari, Yoshimasa Yamano, and Kaoru Kohyama. 2005. Collection of Japanese Texture Terms: Studies on Japanese Texture Terms. *Nippon Shokuhin Kagaku Kogaku Kaishi [Japanese Society for Food Science and Technology]* 52 (8): 337–346.

Hayakawa, Fumiyo, Kana Ioku, Sayuri Akuzawa, Chie Yoneda, Yukari Kazami, Katsuyoshi Nishinari, Yasumasa Baba, and Kaoru Kohyama. 2006. Shitsumon Shihoo Ni Yoru Shoohisha No Tekusuchaa Goi Choosa [Research Survey of Japanese Consumers on Texture Vocabulary]. *Nippon Shokuhin Kagaku*

Kogaku Kaishi [Journal of the Japanese Society for Food Science and Technology] 53 (6): 327–336.
Hayakawa, Fumiyo, Sayuri Akuzawa, Chie Yoneda, Yukari Kazami, Katsuyoshi Nishinari, Yoshihiro Nakamura, Yasumasa Baba, and Kaoru Kohyama. 2007. Seebetu/Nenreebetu/Tiikibetu Ni Mita Nihongo Tekusutyaa Yoogo No Ninti Zyookyoo [Recognition of Japanese Texture Descriptive Terms According to Gender, Age and Region]. *Nippon Syokuhin Kagaku Koogaku Kaisi [Food Science and Technology Research]* 54 (11): 488–502.
Hines, Caitlin. 1994. "Let Me Call You Sweetheart": The Woman as Dessert Metaphor. In *Cultural Performances: Proceedings of the Third Berkeley Women and Language Conference*, ed. Mary Bucholtz, A.C. Liang, Laurel A. Sutton, and Caitlin Hines, 295–303. Berkeley, CA: Berkeley Women and Language Group.
———. 1999. Rebaking the Pie: The Woman as Dessert Metaphor. In *Reinventing Identities: The Gendered Self in Discourse*, ed. Mary Bucholtz, A.C. Liang, and Laurel A. Sutton, 145–162. Oxford: Oxford University Press.
Horsey, Richard. 1998. *Null Arguments in English Registers*. Thesis, La Trobe University, Bundoora.
Hoshino, Yuko. 2015. Gurume Kiji Ni Okeru Onomatope [Onomatopoeia in Gourmet Articles]. *Japanese Language Education in Europe* 19: 90–94.
Kagitani, Tatsuki, Mao Goto, Junji Watanbe, and Maki Sakamoto. 2014. Sound Symbolic Relationship between Onomatopoeia and Emotional Evaluations in Taste. *Proceedings of the Annual Meeting of the Cognitive Science Society* 36: 2871–2876.
Karatsu, Mariko. 2004. Verbal and Nonverbal Negotiation in Japanese Storytelling. In *Hidden and Open Conflict in Japanese Conversational Interaction*, ed. Polly Szatrowski, 125–161. Tokyo: Kuroshio Shuppan.
———. 2012. *Conversational Storytelling among Japanese Women*. Amsterdam: John Benjamins Publishing. https://benjamins.com/catalog/sin.16.
Kato, Ayumi, Yusuke Fukazawa, and Taketoshi Mori. 2012a. Discovering Pattern of Onomatopoeia Used in Food Reviews. In *The 6th International Conference on Soft Computing and Intelligent Systems, and The 13th International Symposium on Advanced Intelligence Systems*, 502–507. Kobe, Japan. https://doi.org/10.1109/SCIS-ISIS.2012.6505214.
Kato, Ayumi, Yusuke Fukazawa, Tomomasa Sato, and Taketoshi Mori. 2012b. Extraction of Onomatopoeia Used for Foods from Food Reviews and Its Application to Restaurant Search. In Lyon, France. https://doi.org/10.1145/2187980.2188192.
Koike, Chisato. 2001. An Analysis of Shifts in Participation Roles in Japanese Storytelling in Terms of Prosody, Gaze and Body Movements. *Proceedings of*

the Twentyseventh Annual Meeting of the Berkeley Linguistics Society (Vol. 27, pp. 381–392).

———. 2009. *Interaction in Storytelling in Japanese Conversations: An Analysis of Story Recipients' Questions*. Doctoral dissertation, University of California at Los Angeles, Los Angeles, CA, USA.

———. 2010. Ellipsis and Action in a Japanese Joint Storytelling Series: Gaze, Pointing, and Context. In *Storytelling Across Japanese Conversational Genre*, ed. P. Szatrowski, 61–111. Amsterdam: John Benjamins.

Koo, Yeon-Hwa. 2009. An Advertisement and Food: Vocabulary of a Television Commercial in Japan and Korean. *The 3rd Consortium on Global Perspectives in Japanese Studies*.

Kulick, Don, and Bambi Schieffelin. 2004. Language Socialization. In *A Companion to Linguistic Anthropology*, ed. Alessandro Duranti, 349–368. Oxford: Blackwell.

Kuroshima, Satomi. 2010. Another Look at the Service Encounter: Progressivity, Intersubjectivity, and Trust in a Japanese Sushi Restaurant. *Journal of Pragmatics* 42 (3): 856–869. https://doi.org/10.1016/j.pragma.2009.08.009.

———. 2014. The Structural Organization of Ordering and Serving Sushi. In *Language and Food: Verbal and Nonverbal Experiences*, ed. Polly Szatrowski, 54–75. Amsterdam: John Benjamins Publishing. https://benjamins.com/catalog/pbns.238.03kur.

Kusumi, Takashi. 2005. Kokoro de Ajiwau—Mikaku Hyogen o Sasaeru Ninchi No Shikumi [Taste with Mind—The Cognitive Mechanism for Taste Expressions]. In *Aji Kotoba No Sekai [The World of Tasty Expressions]*, ed. Kenichi Seto, Takashi Kusumi, Shigeo Sawai, Tomoko Tsujimoto, Haruhiko Yamaguchi, and Shunsuke Koyama, 88–122. Tokyo: Kaimeisha.

Lakoff, Robin Tolmach. 2006. Identity à La Carte: You Are What You Eat. In *Discourse and Identity*, ed. Anna De Fina, Deborah Schiffrin, and Michael Bamberg, 142–165. Amsterdam: John Benjamins Publishing. https://doi.org/10.1017/CBO9780511584459.008.

Lawson, Nigella. 2001. Creme Brulée. *Nigella.Com*. http://www.nigella.com/recipes/creme-brulee.

———. 2012. Soba Noodles With Sesame Seeds. *Nigella.Com*. http://www.nigella.com/recipes/soba-noodles-with-sesame-seeds.

Lischer, Brian. 2015. 7 Popular Types of Brand Names. Ignyte (blog), October 19. http://www.ignytebrands.com/7-popular-types-of-brand-names/.

Magee, H. Melissa. 2009. Talking About Taste: How the Description of Food Means and Does. Swarthmore College, BA Thesis. http://www.swarthmore.edu/sites/default/files/assets/documents/linguistics/2009_Magee.pdf.

Majid, Asifa, and Stephen C. Levinson. 2011. The Senses in Language and Culture. *The Senses and Society* 6 (1): 5–18. https://doi.org/10.2752/174589311X12893982233551.

Mannion, Niamh. 2016. NINA Capri: Recipes from Italy's Amalfi Coast | Cookbook Review. TheTaste.Ie (blog), July 4. http://thetaste.ie/wp/a-taste-of-ladolce-vita-in-nina-capri-cookbook-review/.

Matsuo, Shoko. 2014. *Studies of Japanese Expressions Describing Food*. Kyoto: Kyoto Prefectural University.

Matwick, Keri, and Kelsi Matwick. 2014. Storytelling and Synthetic Personalization in Television Cooking Shows. *Journal of Pragmatics* 71 (September): 151–159. https://doi.org/10.1016/j.pragma.2014.08.005.

Mulvenna, Catherine, Edward Hubbard, Vilayanur Ramachandran, and Frank Pollick. 2004. The Relationship Between Synaesthesia and Creativity. In *Eleventh Annual Meeting of the Cognitive Neuroscience Society*. San Francisco.

Muto, Ayaka. 2001. Mikaku Keiyoshi "Amai" to "Karai" No Tagi Kozo. [Polysemous Structure of Taste Adjectives Amai and Karai]. *Journal of Japanese Language Teaching* 110: 42–51.

———. 2003. Aji Kotoba No Giongo Gitaigo: Shoku o Onomatope [Onomatopoeia in Taste Expressions: Onomatopoeia for Food]. In *Kotoba Wa Aji o Koeru*, ed. Kenichi Seto, 241–300. Tokyo: Kaimeisha.

Newman, John, ed. 2009. *The Linguistics of Eating and Drinking*. Amsterdam: John Benjamins Publishing. https://benjamins.com/catalog/tsl.84.

Norrick, Neal R. 2011. Conversational Recipe Telling. *Journal of Pragmatics* 43 (11): 2740–2761. https://doi.org/10.1016/j.pragma.2011.04.010.

Ochs, E. 2002. Becoming a Speaker of Culture. In *Language Socialization and Language Acquisition: Ecological Perspectives*, ed. C. Kramsch, 99–120. New York: Continuum Press.

Ochs, Elinor, and Bambi B. Schieffelin. 1982. Language Acquisition and Socialization: Three Developmental Stories and Their Implications. Sociolinguistic Working Paper Number 105. In *Cultural Theory: Essays on Mind, Self and Emotion*, ed. Richard Schweder and Robert A. LeVine. New York: Cambridge University Press. https://eric.ed.gov/?id=ED252065.

Ochs, Elinor, and Merav Shohet. 2006. The Cultural Structuring of Mealtime Socialization. *New Directions for Child and Adolescent Development* 2006 (111): 35–49. https://doi.org/10.1002/cd.154.

Ochs, Elinor, Clotilde Pontecorvo, and Alessandra Fasulo. 1996. Socializing Taste. *Ethnos* 61 (1–2): 7–46. https://doi.org/10.1080/00141844.1996.9981526.

Oliver, Jamie. 2013. *Pork Stuffing | Pork Recipes | Jamie Oliver Recipes*. Jamie Oliver. https://www.jamieoliver.com/recipes/pork-recipes/stuffing/.

Özdamar, Melda. 2012. *The Visual Language of Cook Books*. Izmir: Izmir University of Economics.

Sasamoto, Ryoko, and Rebecca Jackson. 2016. Onomatopoeia–Showing-Word or Saying-Word? Relevance Theory, Lexis, and the Communication of Impressions. *Lingua* 175: 36–53.

Schieffelin, B.B., and E. Ochs. 1986. Language Socialization. *Annual Review of Anthropology* 15 (1): 163–191. https://doi.org/10.1146/annurev.an.15.100186.001115.

Seto, Kenichi. 2003. *Kotoba Wa Aji Wo Koeru: Oishii Hyogen No Tankyu [Words Go Beyond Taste: Exploration of Tasty Expressions]*. Tokyo: Kaimeisha.

Shohet, Merav. 2007. Narrating Anorexia: "Full" and "Struggling" Genres of Recovery. *Ethos* 35 (3): 344–382.

Simner, Julia. 2012. Defining Synaesthesia. *British Journal of Psychology* 103 (1): 1–15. https://doi.org/10.1348/000712610X528305.

Simner, Julia, Catherine Mulvenna, Noam Sagiv, Elias Tsakanikos, Sarah A. Witherby, Christine Fraser, Kirsten Scott, and Jamie Ward. 2006b. Synaesthesia: The Prevalence of Atypical Cross-Modal Experiences. *Perception* 35 (8): 1024–1033.

Simner, Julia, Louise Glover, and Alice Mowat. 2006a. Linguistic Determinants of Word Colouring in Grapheme-Colour Synaesthesia. *Cortex* 42 (2): 281–289. https://doi.org/10.1016/S0010-9452(08)70353-8.

Slater, Nigel. 2018. Nigel Slater's Crisp Aubergines with Chilli Sauce Recipe. *The Guardian*, February 20, sec. Food. https://www.theguardian.com/lifeandstyle/2018/feb/20/nigel-slater-crisp-aubergines-with-chilli-sauce-recipe.

Sneijder, Petra, and Hedwig F.M. Te Molder. 2005. Disputing Taste: Foods Pleasure as an Achievement in Interaction. *Appetite* 45 (1): 51–61. https://doi.org/10.1016/j.appet.2005.03.002.

Sperber, Dan, and Deirdre Wilson. 1986/1995. *Relevance: Communication and Cognition*, 2nd ed. Oxford: Blackwell.

———. 2015. Beyond Speaker's Meaning. *Croatian Journal of Philosophy* 15 (2): 117–149.

Strauss, Susan. 2005. The Linguistic Aestheticization of Food: A Cross-Cultural Look at Food Commercials in Japan, Korea, and the United States. *Journal of Pragmatics, Focus-on Issue: Discourse and Metadiscourse* 37 (9): 1427–1455. https://doi.org/10.1016/j.pragma.2004.12.004.

Streeck, Jürgen, Charles Goodwin, and Curtis LeBaron. 2011. *Embodied Interaction: Language and Body in the Material World*. Cambridge: Cambridge University Press. https://books.google.ie/books?id=g7vqkiguszYC.

Szatrowski, Polly E. 2000a. Relation Between Gaze, Head Nodding and Aizuti "Back Channel" at a Japanese Company Meeting. *Annual Meeting of the Berkeley Linguistics Society* 26 (1): 283–294. https://doi.org/10.3765/bls.v26i1.1154.

———. 2000b. Kyoodoo hatuwa ni okeru sankasya no tatiba to gengo/higengo koodoo no kanren ni tuite [Relation Between Participants Status and Verbal/Nonverbal Behaviour in Co-construction]. *Nihongo Kagaku* 7: 44–69.

———. 2002a. Animeesyon No Sutoorii o Kataru Sai No Wadan to Tyuusin Hatuwa. *Hyoogen Kenkyuu* 76: 33–39.

———. 2002b. Nitibee in Okeru Animeesyon No Sutoorii No Katarikata to Higengo Koodoo No Sooi. In *Soogooteki Nihongo Kyooiku o Motomete*, ed. Osamu Mizutani and Deok-bong Lee, 187–201. Tokyo: Kokusho Kankokai.

———. 2003. Gaze, Head Nodding and Aizuti "Back Channel Utterances" in Information Presenting Activities. *Japanese/Korean Linguistics* 11: 119–132.

———. 2004. *Hidden and Open Conflict in Japanese Conversational Interaction*. Tokyo: Kuroshio Shuppan. https://experts.umn.edu/en/publications/hidden-and-open-conflict-in-japanese-conversational-interaction-2.

———. 2005a. Danwa to Buntai—Kanzyoo Hyooka No Dooteki Na Katee Ni Tuit [Discourse and Style—On the Dynamic Process of Emotion/Evaluation]. In *Hyoogen to Buntai [Expression and Style]*, ed. Akira Nakamura, Mayumi Sakuma, Masaaki Nomura, and Chizuko Komiya, 468–480. Tokyo: Meiji Shoin. https://experts.umn.edu/en/publications/danwa-to-buntai-kanzyoo-hyooka-no-dooteki-na-katee-ni-tuite.

———. 2005b. Zyoohoo Syori, Soogo Sayoo, Danwa Koozoo Kara Mita Tooti to Higengo Koodoo to No Kankee. In *Katudoo to Site No Bun to Hatuwa [Sentences and Utterances as Activities]*, ed. Shuya Kushida, Toshiyuki Sadanobu, and Yasuharu Den, 159–208. Tokyo: Hitsuji Shobo.

———. 2008. Koogi Danwa Ni Okeru Sisee, Muimi Na Doosa, Miburi Ni Tuite. *Hyoogen Kenkyuu* 88: 24–37.

———. 2010a. Koogi No Danwa No Higengo Koodoo. In *Koogi No Danwa No Hyoogen to [Expression and Understanding in University Lectures]*, ed. Mayumi Sakuma, 187–204. Tokyo: Kuroshio Shuppan. https://experts.umn.edu/en/publications/koogi-no-danwa-no-higengo-koodoo.

———., ed. 2010b. *Storytelling Across Japanese Conversational Genre*. Amsterdam: John Benjamins Publishing. https://experts.umn.edu/en/publications/storytelling-across-japanese-conversational-genre.

———. 2013. Tabemono o Hyooka-Suru Sai Ni Motiirareru "Kyakukanteki Hyoogen" to "Syukanteki Hyoogen" Ni Tuite [On the Use of "Subjective" and "Objective" Expressions for Food Assessment in Japanese]. *Kokuritu Kokugo Kenkyuuzyo Ronbunsyuu [NINJAL Research Papers]* 5: 95–120.

———., ed. 2014a. Introduction to Language and Food. In *Language and Food: Verbal and Nonverbal Experiences*, 3–31. Amsterdam: John Benjamins Publishing. https://benjamins.com/catalog/pbns.238.01sza.

———., ed. 2014b. *Language and Food: Verbal and Nonverbal Experiences.* Amsterdam: John Benjamins Publishing.

Takasaki, Midori. 2012. "Bimi" o Imi-Suru Go No Siyoo to Seesa- "Oisii" o Tyuusin Ni- [On Gender Differences and the Use of Words That Mean Bimi (Good Flavor) in Japanese: Focusing on Oisii (Delicious)]. *Otyanomizu Zyosi Daigaku "Zinbun Kagaku Kenkyuu" [Ochanomizu University "Studies in Arts and Culture"]* 8: 55–67.

Tamori, Ikuhiro, and Lawrence Clifford Schourup. 1999. *Onomatope: Keitai-to Imi [Onomatopoeia: Form and Meaning]*. Tokyo: Kuroshio Shuppan.

Uchida, Yuzu, Junpei Watanabe, Jun Yoneyama, and Kenji Araki. 2011. Affect Analysis of Onomatopoeia Sentences Extracted from Blog Entries. In *17th Association for Natural Language Processing Conference*, 332–335.

Ward, Jamie, and Julia Simner. 2003. Lexical-Gustatory Synaesthesia: Linguistic and Conceptual Factors. *Cognition* 89 (3): 237–261.

Ward, Jamie, Julia Simner, and Vivian Auyeung. 2005. A Comparison of Lexical-Gustatory and Grapheme-Colour Synaesthesia. *Cognitive Neuropsychology* 22 (1): 28–41. https://doi.org/10.1080/02643290442000022.

Ward, Jamie, Brett Huckstep, and Elias Tsakanikos. 2006. Sound-Colour Synaesthesia: To What Extent Does It Use Cross-Modal Mechanisms Common to Us All? *Cortex* 42 (2): 264–280. https://doi.org/10.1016/S0010-9452(08)70352-6.

Ward, Jamie, Daisy Thompson-Lake, Roxanne Ely, and Flora Kaminski. 2008. Synaesthesia, Creativity and Art: What Is the Link? *British Journal of Psychology* 99 (1): 127–141. https://doi.org/10.1348/000712607X204164.

Watanabe, Chiemi, and Satoshi Nakamura. 2012. *Onomatoperori: Search Engine for Cooking Recipes by Impression of Their Tastes or Textures Using Japanese Onomatopoeia*. Kobe.

Wharton, Tim. 2009. *Pragmatics and Non-Verbal Communication*. Cambridge: Cambridge University Press. https://doi.org/10.1017/CBO9780511635649.

———. 2010. Recipes: Beyond the Words. *Gastronomica: The Journal of Food and Culture* 10 (4): 67–73.

Yoshikawa, Seiji, Shinya Nishimaru, Toyohisa Tashiro, and Masaki Yoshida. 1970. Collection and Classification of Words for Description of Food Texture III: Classification by Multivariate Analysis. *Journal of Texture Studies* 1 (November): 452–463. https://doi.org/10.1111/j.1745-4603.1970.tb00744.x.

Zwicky, Ann D., and Arnold M. Zwicky. 1980. America's National Dish: The Style of Restaurant Menus. *American Speech* 55 (2): 83. https://doi.org/10.2307/3050498.

6

Onomatopoeia and the Showing–Saying of Japanese Culture

So far, I have presented a relevance-theoretic account of the role of onomatopoeia in communication. I have also discussed the lexical aspect of onomatopoeia using notions particularly from relevance-theoretic lexical pragmatics: lexical adjustment, and ad hoc concept formation. In this chapter, I turn to a culture-specific aspect of onomatopoeia, with a particular focus on onomatopoeia used in Japanese manga.

1 Japanese Culture and Visual Nature

It is often claimed that Japanese culture is highly visually oriented. For example, Kitamura (1987, 144) states that "of the five senses, sight is most important in Japan". MacWilliams (2008) also describes Japan as an instance of what Susan Sontag (1999, 80) calls "an image world", which describes a society among whose "chief activities is producing and consuming images, when images that have extraordinary powers to determine our demands upon reality and are themselves coveted substitutes for first-hand experience".

R. Sasamoto, *Onomatopoeia and Relevance*, Palgrave Studies in Sound,
https://doi.org/10.1007/978-3-030-26318-8_6

There are a number of reasons for such claims. Cooper-Chen (2010, 10) lists instances of the visual nature of Japanese culture, such as the "ideographic writing system, the rich tradition of painting and wood-block painting, the variety of ceramic and fabric designs, the importance of artistic food and flower arrangements, the attention to packaging", and claims that such instances underscore the observation that Japanese culture is highly visually oriented. Gravett (2004, 18) also mentions how Japanese people have a "long-standing appetite for pictorial art, which [...] at times gets close to the spirit of the comic". The highly multimodal nature of Japanese media, including the use of textual inserts on TV programmes, as reported by Shiota (2003), O'Hagan (2010), Shitara (2011), Sasamoto (2014), and Sasamoto et al. (2017), also adds to the argument that Japanese culture is visually oriented.

Of course, the height of multimodal media with a particular focus on visual elements is manga. As MacWilliams (2008, 6) explains, manga is a hybrid representation of visual and verbal elements, and the narrative is told not just verbally but also through visuals. Manga is one of Japan's biggest cultural exports and has a global fan base. However, unlike the common conception, manga is not just an instance of popular culture. Rather, it has its roots firmly cemented in Japanese traditional culture. The origin of manga is said to be the *Choju Giga* scroll from the eleventh century,[1] and the term manga was said to be coined by Katsushika Hokusai, with his famous *Hokusai Manga*, the encyclopaedia for whimsical drawings.

Some may argue that manga is not particular to Japanese culture, as Western cultures also have a long history of comics. However, Berndt (2011) maintains that comics and manga are totally different, in terms of publishing, distribution, and consumption. Unlike the perception of manga abroad, where it is seen as a specific genre only for fan readers, manga in Japan is just one type of media and enjoys huge popularity across a diverse readership (Berndt 2011). The publication of weekly manga peaked in the 1990s, before the era of smartphones, when manga industry earned "three times the revenue of the domestic film industry, and accounted for about 38 percent of all published matter, and 22 percent of all published revenue" (Kinsella 1999, 567). Furthermore, it was recorded that "there were 12 mag-

[1] Some scholars (e.g. MacWilliams 2008) argue otherwise.

azines with a circulation of 1 million or above, and approximately 50 magazines with circulations between 150,000 and 1 million" (ibid.). Schodt (1996) also cites a study conducted by the Research Institute for Publications, which shows that sales of manga in 1995 accounted for approximately 40% of all the books and magazines sold in Japan. Even after smartphones became a household item and book sales in general started to decline, manga sales were 481 billion in 2006 (Schodt 1996).

In addition, manga in Japan is not a specific genre of comics. Unlike US comics, which are expected to have an extraordinary hero doing extraordinary stuff, "the real triumph of Japanese manga lies in their celebration of the ordinary" (Schodt 1996, 28). Furthermore, manga is part of an everyday media platform, and we see manga-style media products everywhere, including adverts, recipes, and even textbooks. The Ministry of Education, Culture, Sports, Science and Technology is considering introducing manga into the school curriculum as a new way of improving literacy. Indeed, there has been some debate on the definition and genres of manga (see Gravett 2004). However, it is beyond the scope of this chapter to discuss the characteristics of manga. Instead, the focus of this study is the use of onomatopoeia, which is said to be one of the most defining features of manga, having both verbal and non-verbal elements.

2 Manga and Onomatopoeia: Interaction of Image and Language

Onomatopoeia is said to be one of the characteristics that distinguishes Japanese manga from other types of comics. Onomatopoeia is highly prevalent in manga and is an integral part of the aesthetics. Schwartz and Ribinstein-Ávila (2006) argue that onomatopoeia is an integral part of the storytelling of manga and creates dynamics and atmosphere. Of course, this is not to say that onomatopoeia does not appear in Western comics. However, in contrast to the insertion of onomatopoeia in Western comics, where it is provided very much as an add-on in a particular cell, onomatopoeia in manga is often presented in a highly stylised font and as part of the aesthetics.

Pratha et al. (2016) conducted a corpus analysis of onomatopoeia, which indicated that there are differences in the use of onomatopoeia between American comics and Japanese manga. Schodt (1983) also describes the difference between onomatopoeia in manga and comics, in terms of the number and variety of expressions. According to Schodt (1983, 23), onomatopoeia in American comics, despite its long tradition, is limited to that describing "explosions (BAROOM!), fists impacting on jaws (POW!), and machine guns zapping away at Commies (BUDABUDABUDA)", while Japanese manga uses a wider range of onomatopoeic expressions, starting from the slurping of noodles, types of rain, the lighting of a cigarette lighter, and even silence. It is well known that Japanese has an extremely rich repertoire of onomatopoeia, and this is also true for onomatopoeia in manga. A research team led by Cathy Sell and Darah Pasfield-Neofitou compiled 769 onomatopoeic expressions from professionally published manga and their 1330 translation pairings (Manga Studies SFX Glossary).[2] In terms of the stylistics of onomatopoeia in comics, Guynes (2014, 61, original emphasis) characterises onomatopoeia in comics this way: "Brightly colored block letters reading "*pow*" or "*bang*" or "*whap*" are overlaid on separately, but equally brightly colored geometric designs with jagged edges radiating outward." In contrast, onomatopoeia in manga is often hand drawn, as part of the aesthetics, rather than being overlaid separately, and is presented not only in block letters but in a variety of forms, depending on what the artist intends to depict in a given scene.

Despite the difference in how they appear on the page, onomatopoeia in comics and manga have something in common. As we reported elsewhere (Rohan et al. 2018), scholars often consider onomatopoeia in manga and comics as sound effects (SFX; cf. Eisner 2008; Hyuga 1986; McCloud 1993; Cohn 2014a; Guynes 2014), such as explosions or punches, and vocalisations, such as coughs and grunts (Guynes 2014). For example, drawing on Cohn's (2014b) Visual Narrative Structure, Guynes (2014, 70) analyses the use of onomatopoeia in comics and argues that "comic book onomatopoeia embody all meaning-making modes of Pierce's thirdness, and as signs they indexically suggest the sounds in the real world which they purport (by symbolism) to signify". Similarly, McCloud (2006) argues that onomatopoeia is often depicted

[2] Available at http://mangastudies.com/sfx/about.php. Accessed 18 May 2018.

in highly stylised fonts, distinct from the standard typeface used in comics (cf. narrative and speech bubbles). As McCloud argues, the use of such stylised fonts enables the creator to emphasise a particular sensory experience and the creator's intention:

> Thanks to film and television, we've gotten used to stories that continuously use **sight** and **sound** and offer rich, immersive experiences. But as comics creators, if we want to **reproduce** that kind of experience, we need to do it using only **one** sense. Words play an important role in comics by **bridging** that gap. They give **voice** to our characters, allow us to describe all **five** senses—and in the case of **sound effects**, they graphically **become** what they describe—BANG!—and give readers a rare chance to **listen**—with their **eyes**.
>
> (McCloud 2006, 146, original emphasis)

Natsume (2013), who is a manga artist himself, reports how onomatopoeia was already being discussed in the 1960s. According to Natsume, Kusamori (1968, cited in Natsume 2013) describes onomatopoeia as "a means to communicate matters to the audience in a sensual manner" (Natsume 2013, 218, my translation). As stated by Kusamori (1968, cited in Natsume 2013), manga in onomatopoeia functions in the following four ways: (i) as mimetics to increase realism; (ii) as mimetics that exceed realistic description; (iii) as psychological and symbolic mimetics; and (iv) as humorous mimetics. As Natsume (2013) shows, each type of onomatopoeia characterises the particular era or genre of manga. However, the point is that onomatopoeia played a huge role in the development of manga in Japan. Natsume (2013) further argues that the nature of Japanese (and other Asian) drawing means that realism is not so strictly required and the relationship between images and characters is closer than that of Western aesthetics. That is, according to Natsume (2013), the key to the prevalence of onomatopoeia in manga is the nature of manga aesthetics, which exhibits less autonomy than the Western art tradition of realism.

Gravett (2004, 20) also explains how the nature of Japanese writing means that the printing of words and images was easier than it was in the West. He argues:

> [I]n books using the Roman alphabet [...] blocks of text were composed employing a 26 letter alphabet of moveable type; any woodcut illustration would need to be made separatel[y], so early publishers of English books and broadsheets preferred to save trouble and expense by keep pictures to a minimum, or drop them entirely. In Japan, however, moveable type took longer to take hold because the language requires so many more letters and characters. Thus it was easier to cut the words onto the same wood block as the illustration. The upshot of this was that from the start text could be smoothly incorporated within images and the two elements conceived, printed and read together, as in the basic system of comics.

So, Natsume's (2013) analysis of onomatopoeia in manga, and Gravett's account of the number of characters in the alphabet, do indeed provide some explanation as to how onomatopoeia became so prevalent in manga. However, for a study that is concerned with the role of onomatopoeia in communication, the question is what onomatopoeia does in manga. A number of scholars also address this. For example, Jun Sasamoto (2000, 16, cited in Natsume 2013, my translation) explains the creativity of onomatopoeia as follows:

> The size of characters corresponds to the size/degree of the sound or feeling expressed, and the change in the shape is designed to synaesthetically communicate the tone, emotional nuance, or the sense of speed. That is, [the visual element of onomatopoeia] supplements and emphasises what the sound onomatopoeia as a word cannot fully express.

McCloud (2006, 146–147) also shows techniques for comic artists to use SFX effectively, which includes utilising size, boldness, tilt, and exclamation marks for loudness; roughness, waviness, sharpness, and fuzziness for timbre (or the quality of the sound); typeface and shapes to refer to the source of the sound (e.g. a water-like shape of font to express the splash of water), and so on.

Similarly, Petersen (2009), in his discussion of sound in comics, argues that Japanese manga exploits sound more effectively than any other comics, due to their scripts; their suitability to be used as part of aesthetics; Japan's history of storytelling; and the rich repertoire of onomatopoeia in the Japanese language. On the function of onomatopoeia in manga, Petersen (2009, 163) argues:

> [Japanese] Comics utilize word/pictures and pictured/words, where the way something is written visually informs sound qualities, in narrative action. Such exaggerated onomatopoeic words commonly appear in comics as hybrid word/pictures, which convey the essence of lived sensations by using the sound-like experience to fuse the sign/icon into a single sensation.

Petersen further argues that onomatopoeia in manga takes a supportive role rather than creating the narrative itself:

> The sounds do not create the dramatic action: rather they provide time for the intensity of the action to develop. The presence of the sound gives force and dimension to the dramatic action.
>
> (Petersen 2009, 166)

Tajima (2006, 202) seems to agree with Petersen's idea that onomatopoeia provides time and intensity for the action described and argues that onomatopoeia in manga creates the effect for the reader watching a film or a TV programme. Itoi (1989) and Fukuma (1993) also mention the supportive role of onomatopoeia in manga, such as adding a sense of continuity to the described action. Morooka (2010) also argues that the use of onomatopoeia in manga adds depth by offering another layer of expression to the scene described.

So, there seems to be a consensus that onomatopoeia in manga plays a significant supporting role in enhancing what is communicated through the narrative delivered via speech and images. That is, it seems to be the case that onomatopoeia in manga, as has been discussed, bridges images and language and adds nuances and emotions. As is often pointed out, the key characteristic of onomatopoeia in manga is its presentation, which is frequently in a highly stylised font with varying sizes. In some extreme cases, it is so highly stylised that it is almost impossible to read—even for native speakers. Such presentation blurs the very essence of onomatopoeia in manga. For example, Natsume (1997, 89, my translation) argues that onomatopoeia in manga is "words but the image at the same time" when it is drawn by hand in the image. Natsume (1997, 16–17) discusses onomatopoeia in manga as follows:

> Onomatopoeia in manga has an aspect of being a verbal language and an aspect of being letters that are drawn = pictures. That is, [onomatopoeia in manga] is a phonological symbol as well as an image symbol that require a special treatment. [...] I therefore coined the term "sound metaphor" for onomatopoeia in manga.

This suggests that for a theory to account for the role of onomatopoeia in manga, it needs to be able to account for the role of onomatopoeia, both as verbal and as non-verbal inputs, in communication. The aim of this chapter is therefore to present a relevance-theoretic account of onomatopoeia in manga.

3 Relevance-Theoretic Approach to Onomatopoeia in Manga

Throughout this study, I have been arguing that from the perspective of relevance theory, onomatopoeia has both verbal and non-verbal elements and is located on the *showing–saying* continuum. Onomatopoeia provides both direct and indirect evidence for inference (Sasamoto and Jackson 2016). I have argued that onomatopoeia's *showing* element provides direct evidence for communication, bringing about a similar impression from a (different) sensory domain, while some onomatopoeia also has *saying* elements, which provide indirect (coded) evidence for communication. Onomatopoeia's direct evidence provides quick and dirty access to non-propositional effects that are based on a perceptual resemblance to sensory experience. That is, onomatopoeia is humans' exploitation of connections over (different) sensory domains, allowing for the sharing of impressions, in addition to propositional contents (in some cases) communicated via indirect evidence.

This analysis enables us to account for how the use of onomatopoeia in manga communicates such nuances or emotions. In what follows, I will show how the relevance-theoretic account predicts the role of onomatopoeia in manga, which is to communicate intangible, nebulous impressions beyond their prescribed verbal meaning.

3.1 Non-verbal Communication

In relevance theory, it is generally acknowledged that the interpretation process is triggered by *any* ostensive communicative behaviour, including naturally occurring behaviours (Sperber and Wilson 1986/1995; Wharton 2009; Scott 2017). Ostensive communicative behaviour involves any communicative stimuli that are deliberately and openly shown to an audience. That is, the hearer's interpretation process will begin, whether the communicator uses a linguistically encoded stimulus (an utterance), a gesture, a facial expression, or even just an eye movement, as long as the communicator presents their intention to communicate deliberately and openly. This allows for non-verbal (non-coded) communication to be treated in the same way as linguistic input, which is perfect for the analysis of onomatopoeia in manga.

That said, so far the focus of relevance-theoretic work has largely been on verbal communication. However, in recent years the link between verbal and non-verbal features of communication has begun to attract attention, such as Wharton's (2009) work on non-verbal communication and Yus's (2009) work on multimodal metaphors. What is particularly relevant to onomatopoeia in manga, among these relevance-theoretic works on non-verbal communication, is Scott and Jackson's (forthcoming) work on typography. Scott and Jackson (forthcoming) take a relevance-theoretic approach and examine a range of typographical features, particularly capitalisation and italicisation. They argue that the typographical choices are driven by the author's communicative intentions and considerations of relevance, guiding the reader to reach the intended interpretation. See examples (1) to (3):

(1) Andrew threw to Patrick and then *he* threw to Matthew.
(2) …duly rehearsing a small ceremony known in high ecclesiastical life as the *nolo episcopari…* (George Eliot, *Silas Marner*, p. 73).
(3) Oh—you're *Jor*dan Baker (F. Scott Fitzgerald, *The Great Gatsby*, p. 23).

Examples from Scott and Jackson (forthcoming, 2)

According to Scott and Jackson (forthcoming, 2–3), the italicisation of the pronoun *he* in example (1) influences the assignment of referent from Andrew to Patrick, while in (2) the italicisation of *nolo episcopari* marks a Latin term in a text written in English. Similarly, the italicisation of the part of the name in (3) creates "a sense of contrast and emphasis respectively, and [is] communicating the speaker's attitude or emotion".

Scott and Jackson (forthcoming) cite House (2006) and further show that capitalisation can also be used in this way:

(4) Jenny teaches violin
(5) JENNY teaches violin
(6) Jenny TEACHES violin
(7) Jenny teaches VIOLIN

As Scott and Jackson (forthcoming) argue, all four utterances have the same explicit, truth-conditional meaning. That is, all utterances are true if and only if Jenny teaches violin. The differences between them, according to the authors, come from the contrast conveyed by each in terms of the state of affairs described in each utterance and another possible state of affairs. That is, the capitalisation of Jenny in (5) triggers the search for relevance in the fact that it is Jenny, not anyone else, who teaches violin, while in (6) the contrast the reader is encouraged to identify is related to the fact that Jenny teaches, not plays or learns, violin. Similarly, in (7), the capitalisation of "violin" encourages the reader to search for relevance in the fact that it is the violin, and not another instrument, that Jenny teaches.

Drawing on Wharton's (2009) work on prosody and gesture, Scott and Jackson (forthcoming) describe parallels between typographical features such as unexpected changes in letter form in written utterances, and capitalisation or italicisation with aspects of prosody in spoken utterance, and they suggest that typography, like aspects of prosody, can be explained as functioning procedurally.[3] That is, certain uses of typography can be seen

[3] Note that Scott and Jackson (forthcoming) explicitly state that they do not mean to suggest that typography *encodes* a procedure, as "the effect that the features have on interpretation is context specific and both the triggering process and the effects that are generated are relative to the discourse context of the utterance, widely construed. It is not the presence or absence of a feature that

as, in the way that Wharton explains the role of prosody, "guid[ing] the utterance interpretation process by altering the salience of possible interpretations of utterances" (Wharton 2009, 141). Typographical features such as capitalisation or italicisation act as highlighting devices that trigger the search for extra effects by drawing the reader's attention, while also guiding the reader to follow the path of least effort until they reach the intended interpretation, which justifies the extra processing effort incurred by the unexpected typographical change.

Note that not all typographical features deliver the message with the same degree of determinateness. Sometimes, certain typographical uses might have become conventionalised, such as the use of italics for a foreign word in a text, or block capitals to mark shouting. At other times, however, certain uses of typographical features are context dependent and will be much less determinate, and hence only evoke impressions rather than communicating some concrete proposition. Indeed, according to Sperber and Wilson (1986/1995, 2015), communicated meaning can vary in terms of how determinate the message is. For example, in (8) the communicated message is highly determinate, while the metaphor in (9) communicates a range of weaker propositions that amount only to something highly intangible and indeterminate:

(8) Alfie: What time is your appointment?
Nathan: 3 p.m.

(9) [Students talking about their final exam]
The question on truth-conditional semantics killed me.

Sperber and Wilson (2015) further illustrate that *showing* can also be more or less determinate. For example, Sperber and Wilson (2015, 124) explain that pointing at a clock which shows the time of 5 p.m., when asked for the time, would be a case of determinate *showing*. In contrast, pointing at an approaching black cloud would be a case of less determinate *showing*, which communicates something less determinate than pointing at a clock to show the time.

guides the interpretation but the confirmation or disconfirmation of expectation about how an utterance will be produced" (Scott and Jackson forthcoming, 37).

3.2 The Role of Onomatopoeia in Manga

The notion of the determinate–indeterminate continuum is particularly interesting for the analysis of onomatopoeia in manga. The interpretation of typographical features used in manga, or the effects to which the use of typographical features gives rise, are not conventionalised, and only amount to something that is highly indeterminate and intangible. The question, now, is how typographical features of onomatopoeia in manga give rise to such effects.

As noted elsewhere, onomatopoeia in manga are often presented in a highly stylised manner and frequently blend in with the aesthetics of the page (Rohan et al. 2018). This has led scholars to claim that onomatopoeia is both verbal (linguistic) and non-verbal (aesthetic, or part of an image). This suggests that in addition to carrying both *showing* and *saying* elements as onomatopoeic expressions, onomatopoeia in manga bear further visual elements via their visual representation. That is, onomatopoeia in manga add another layer of meaning via their visual representation, such as font, size, and tilt, which provides further direct evidence for communicating the intended interpretation, in addition to the *showing* aspect (= communicated via the direct evidence for communication) and the *saying* aspect (= what is coded by the expression) of the expression. It would then seem reasonable to suggest that onomatopoeia in manga involves a further *showing* element via its visual presentation. See example (10) (Fig. 6.1):

(10) Onepanman [One Punch Man]

This page depicts a scene in which a huge meteor is approaching. In this example, a number of onomatopoeic expressions are used in different styles. At the top of the right panel, the onomatopoeic expression (a) *woooo* (ウウウウ, siren noise) is presented in a thick black, jagged typeface. This represents the warning siren, which also is marked in the speech bubble nearby that says *Emergency Evacuation Warning* [緊急避難警報]. In the same panel, towards the bottom of the same cell, four onomatopoeic expressions are used: (b) *popoo* (ププー, beep), (c) *paa* (パー, beep), (d) *kya-* (キャー, scream), and (e) *wah* (ワー, scream), but this

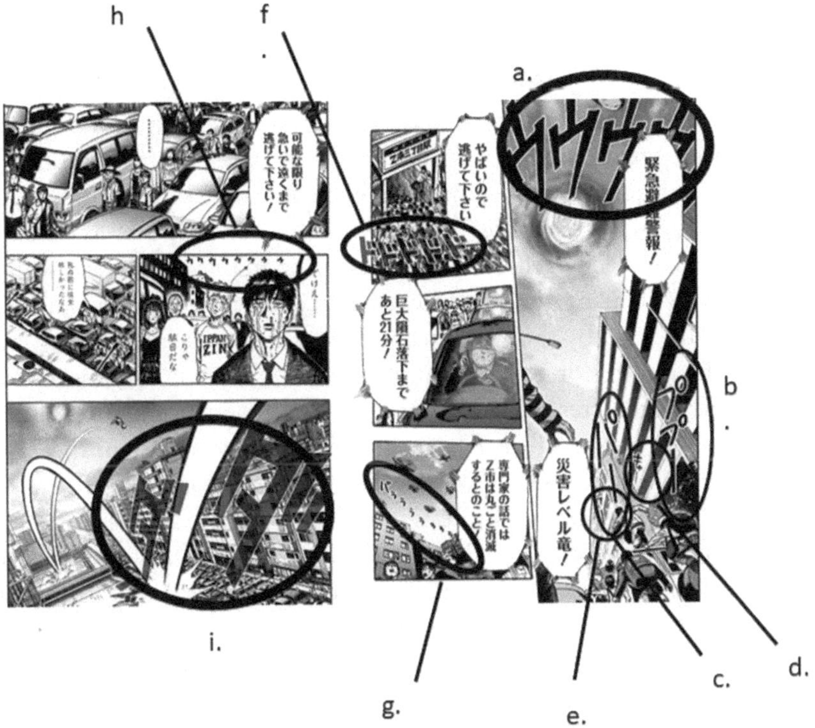

Fig. 6.1 Onomatopoeia in *shonen* (boys) manga, *Onepanman* [One Punch Man] 24 ©ONE, Yusuke Murata/SHUEISHA (https://tonarinoyj.jp/episode/13932016480028985360)

time in a rounder and thinner line. The middle top cell shows (f) *do do do do do* (ドドドドド, thunderous noise) in half-transparent grey, in a thick, square design. The bottom panel in the middle also has an onomatopoeia (g) *barararara* (バラララ, the noise of the approaching helicopter), but this time in a rounder, thinner typeface. The size of the letters increases from right to left, as if to show the sense of perspective of the approaching helicopter. It would also indicate the noise level of the helicopter. It is interesting to observe that the middle of the left panel has exactly the same onomatopoeia as the top right, (h) *woooo* (ウウウウ, siren noise), but this time in a rounder and smaller typeface. Finally, the bottom panel on the left shows two large letters (i) *dada (*ダダ, heavy impact noise), in a half-transparent, square typeface, tilted in alignment with the parabolic line.

As discussed by scholars such as Natsume (1997), it is evident that the different presentation of onomatopoeia creates a different atmosphere. For example, the square shape of *dada* (ダダ, heavy impact noise) in (i), which is also tilted along the motion line, depicts the heaviness of impact of the meteor. The visual presentation of this onomatopoeic expression enhances what is delivered by the expression as an onomatopoeia, which has the *saying* element of being a heavy impact sound communicated via what is represented by the kana script ダ (da), as well as the *showing* element of the impression of such a heavy impact, which is communicated via the perceptual resemblance between the experience of the heavy impact and the sound of the expression delivered by the use of the kana script. However, the use of a grey-scale, half-transparent presentation somewhat gives rise to the impression of being distant and unreal. That is, the visual presentation of the onomatopoeia adds further direct evidence of the impression of the scene. This is particularly noticeable when compared with the same onomatopoeia of the siren *woooo* (ウウウウ, siren noise) in (a) and (h). As an onomatopoeic expression, *woooo* (ウウウウ, siren noise) in both (a) and (h) involves the *saying* element of being a noise of a loud siren represented by the kana script ウ as well as the *showing* aspect, which allows the author to share the impression of the siren via the perceptual resemblance between the noise of the siren and the sound of the word. This provides the direct evidence for communication. The difference between *woooo* (ウウウウ, siren noise) in (a) and *woooo* (ウウウウ, siren noise) in (h) is the visual presentation of the expression. When *woooo* (ウウウウ, siren noise) in (a) and the same onomatopoeia in (h) are compared, it is clear that *woooo* (a) depicts an overwhelming sense of emergency accompanied by the loud ringing of the warning siren, while the use of *woooo* (ウウウウ, siren noise) in (h) does not have such urgency. This is because *woooo* in (h) shows that while the siren is still sounding, it is much more distant, whether physically or psychologically, with the focus of the panel being more on the reaction of passers-by, aghast at the scale of the event. What this suggests is that different atmospheres can be communicated by using the same onomatopoeia but presenting them in a different visual design, as the size or the thickness of the line corresponds to the impression of the scene. That is, the visual presentation of onomatopoeia in manga provides further direct evidence for communication, sharing the impression of the experience via its perceptual resemblance.

Note that this does not mean that the thickness of the line or other visual presentation *encodes* distance or intensity. As Scott and Jackson (forthcoming, 37) argue in their analysis of typographical features, the effect communicated by the use of certain visual features is context specific and can only be delivered relative to the context of the utterance. The point is that the typographical features with which the onomatopoeia is presented will trigger readers to search for extra effects, while guiding them to an intended interpretation. Typographical features are visual inputs by nature and arguably provide direct, rather than indirect (coded), evidence for communication. The extra effects that typographical features produce are highly stylistic and indeterminate and are bound (and relative) to the context of the discourse.

This example is from a specific genre of manga (*shonen* manga), where onomatopoeia is particularly widespread. This is probably because the theme of *shonen* manga is often related to battles or sports, where physical contact is prevalent. However, this does not mean that the use of onomatopoeia is limited to *shonen* manga. Let us now examine the use of onomatopoeia[4] in different genres.

Example (11) is a scene from a manga-style train journey review taken from a children's train magazine (Fig. 6.2). This magazine aims to educate children about trains in Japan (and sometimes abroad), contributing to a great fan base for trains in Japan. In every monthly volume, these rabbit-like characters travel on different Japanese train lines, reviewing local attractions and trains. Here, two panels are presented next to each other, contrasting the scene inside the train (left) and outside (right). The speech bubble on the right panel says *with Seishun 18 Ticket*, the speech bubble on the left panel says *being on the train for six hours*. This image contains three uses of onomatopoeia. In the left panel, onomatopoeia (a) すやすや (*suyasuya*, a state of being asleep) and (b) ぐ andがー (*gugaa*, snoring)

[4] Analysis of examples (11) to (14) first appeared in Rohan et al. (2018).

Fig. 6.2 Manga panels of train and passengers with onomatopoeia, *Tetsuomo!* (2012). Neko Publishing. Vol 60, p. 34

are used, while on the right is the onomatopoeia (c) ゴー (*goo*, roaring noise).

(11) Train and passengers

As predicted by the analysis of onomatopoeia presented in this study, the use of these onomatopoeia allows for the readers to experience what it would be like to be on board, as well as to understand the contrast between the interior and exterior of the train. On the left panel, which depicts the scene inside the train, rabbits are depicted as sleeping—one snoring (with the onomatopoeia *gugaa* in (b)) and the other silently (with the onomatopoeia *suyasuya* in (a)). These onomatopoeic expressions, via the sound delivered by the use of kana script, communicate that the rabbits are dribbling as they sleep, not because of hunger.

Suyasuya is a highly established onomatopoeic expression to describe the state of being asleep, while *gugaa* is not sufficiently established as an onomatopoeic word to become a dictionary entry. However, both *gu* and *ga* are often used when expressing a snoring sound, and the nature of Japanese kana writing allows for a flexible combination of syllables.

Therefore, the combination of these syllables can still deliver sound, regardless of the lexicality of the expression. This way, the linguistic (coded) element of both onomatopoeic expressions represents the state of being asleep (*suyasuya*) and the manner of snoring (*gugaa*), telling readers what is happening. At the same time, the *showing* element of onomatopoeia provides direct evidence for communication, allowing for sharing the impression of the state they are in—the manner in which they are breathing, the kind of noise they are making, and so on. In addition to the *showing* element of the onomatopoeic expressions, the visual presentation of these onomatopoeia also *shows*, or gives direct evidence for the communication of, the further impression of the rabbit on the right sleeping more loudly via the onomatopoeia *gugaa*, which is presented in a slightly bolder and bigger typeface.

Similarly, on the right panel, the use of the onomatopoeia *goo* in (c), which is often used to depict a loud roaring sound, indicates that the train is running very fast while making a loud noise. In this example, the *saying* element of onomatopoeia communicates that the train is running with a loud noise, while the *showing* element of the expression provides direct evidence for the impression of the manner in which this train is running. In contrast, the onomatopoeia *goo*, via its *showing* element, communicates how it must feel to experience such noise (and speed). In addition, *goo* is presented in a thick, chunky, bold typeface in red. This additional aesthetic feature triggers the communication of further non-propositional effects of something being forcefully fast and loud. Furthermore, the difference in style between the typographical features of these expressions triggers the search for extra effects, such as the reinforcement of the stark contrast between inside and outside, allowing for the sharing of impressions. That is, the use of these onomatopoeia guides the readers to recover an intended interpretation; that is, the impression of the experience of how the train itself is running with such a loud noise from outside, while it is quiet enough inside for the rabbits to fall asleep.

The next example (12) is taken from a parenting magazine popular with new parents. This manga is a supplement to the main magazine and aims to share information and experience about parenting. This manga is presented as the author/artist's diary in a very humorous manner (Fig. 6.3).

Fig. 6.3 Panels from a manga in a Japanese parenting magazine, *Hiyoko Club* (2010) Comic vol. 74. Benesse Corporation, issue no. 6, magazine no. 200, 208

(12) Parenting magazine

In this example the two panels are placed side by side, showing a number of new-born babies screaming, while tired nurses are talking to each other, asking "Why do babies all cry when one starts crying?" and saying "It's a mystery" in speech bubbles on the left panel. The text box on the top right says "I won't be left behind", which presumably is an interpretation of what the crying babies are thinking. In the space to the left of these speech bubbles, nurses and cots are filled with onomatopoeic expressions describing for the readers the kind of noise the babies are making. The onomatopoeia include ぐあああ *guaaa*, きあああ *kiaaa*, びえええ *bieee*, and ぎゃおーす *gyaoosu*.

The sheer number and combination of onomatopoeia communicate what it must be like to be in such a space. The *saying* element of onomatopoeia, via its coded element (delivered by kana characters with their sound), provides indirect evidence for communication, while the *showing* element shares the impression of a particular crying behaviour. For example, *bieee* conveys the impression of a high-pitched noise, a shriek, a wail, a longer sound, while *kiaaa* gives the impression of a long, hysterical scream. *Giaa* is something like a forceful grunt of complaint, while *guaaa* gives the impression of screaming, gulping, or clenching up.

In addition, as with examples (10) and (11), the onomatopoeia is presented in a highly stylised typeface. The use of a large and heavy font, which seems to be written with brush and ink, provides further direct evidence for communication, conveying the emotional state one would experience when surrounded by screaming babies—uncontrolled, raw, organic chaos. That is, the sharing of the nurses' experiences in the maternity ward is achieved through the use of onomatopoeia, which works as a tool for *saying* that the babies are crying, as well as *showing* the manner of their crying. Readers will be guided by the use of onomatopoeia (and its particular presentation) to access existing cognitive assumptions and to recover intended effects.

Example (13) is taken from a manga-style advertisement for a famous seafood restaurant. In the two top panels, the restaurant manager is making a recommendation from the menu. The middle two panels depict the scene when the food is delivered, with some explanation of the menu given in speech bubbles (8) and (9). The panel at the bottom right depicts the scene where the restaurant manager explains how to eat the hotpot, while the panel on the bottom left shows the customers' excited faces when the dish is ready to eat.[5] Let us focus on the onomatopoeia located near the food in the bottom two panels (Fig. 6.4):

(13) Manga-style advert

In the panel at the bottom of this example, (a) *howaa* is used. This expression is often used to denote the state of something that is hot with steam. It does not particularly refer to the sound of the steam, but denotes how the steam comes up and spreads. That is, the *saying* element of *howaa* in (a) tells readers what is happening, since it provides indirect (coded) evidence for the communication of hot steam rising from the dish. The sound (syllables) delivered by the use of the kana script delivers what would be closest to the sense, if it were sound. At the same time, the *showing* element of this onomatopoeia communicates the impression of the aromatic steam rising from the hotpot.

[5] In this example, speech bubbles are numbered to indicate the reading direction, and the translation of the text appearing in each speech bubble is given in the figure note.

Fig. 6.4 Panels from a manga-style advert, Manga Kani Doraku (2016, own translation) (1) All looks so tasty!; (2) Sashimi looks good, and so does chawan-mushi (savoury steamed custard); (3) Do you have something like a hotpot dish for family?; (4) If you prefer a hotpot, how about "kani-suki" (crab hotpot), which is our signature dish?; (5) Well, we'll have it then; (6) Sorry for the wait!; (7) Wow, it's transparent and clean!; (8) (They say) the soup is based on white-soy sauce, which hasn't changed since the opening of this restaurant!; (9) It has long tradition!; (10) First, we add vegetables that would draw out sweetness; (11) Crab will be done a couple of minutes after adding to the pot. Wow that looks brilliant!; (12) This smells great!; (13) Bon appétit!

As shown with *suyasuya* in the train example and *howaa* here, the use of onomatopoeia is not limited to the aural sense. Indeed, I showed in Chap. 3 that onomatopoeia can communicate a multimodal experience. *Gutsugutsu* in (b), which is an onomatopoeia appearing in the panel at the bottom right, clearly illustrates this. The onomatopoeia *gutsugutsu* is often used to denote the state of something being cooked in sauce (such as a stew or curry). Not only does the impression refer to the sound of the cooking, it also denotes other aspects of cooking, such as a visual element (the slight movement of the food while being cooked in the pot) or the heat that is rising from the pot. That is, *gutsugutsu* describes the culinary process not just via the aural sense, but also from other senses. The kana syllables that represent *gutsugutsu* (or the *saying* element) tell the readers what is happening (the dish being cooked is both making a noise and moving slightly around in the pot, with the rising heat), while the *showing* element provides direct evidence for the particular kind of noise and movement. What is also interesting is the position of the onomatopoeia *gutsugutsu* in the panel. It is positioned just above the dish, as if to cover the pot. This positioning of the onomatopoeia contributes to the communication of the way in which the dish is being cooked evenly throughout the pot, further giving the impression of the appetising state of the hotpot.

The panel on the left has two further onomatopoeia: (c) *gokuri* (below the face of the customer on the left) and (d) *gubi* (below the face of the customer on the right). The use of these expressions *says* what is happening—both customers swallow their saliva—while also *showing* how excited they are. As discussed elsewhere (Rohan et al. 2018), the manga-style advert is particularly effective, as the use of onomatopoeia as well as other aesthetic and linguistic elements in the manga "communicates the quality and atmosphere of the restaurant (and of the food served in it), thereby enhancing the appeal of the dish and thus reinforcing the evaluative claim of this particular advert" (Rohan et al. 2018, 236).

Next, let us see the use of onomatopoeia that describes the emotional state of characters, rather than the state or sound of affairs/action. Such onomatopoeia are often called psychomimes in Japanese linguistics and involve expressions such as *dokidoki* or *dokin* (nervous, based on the pounding heart) or *gakkuri* (disappointed). These expressions are frequently used in the *shojo* manga genre. *Shojo* manga targets young female readers and topics typically involve school life and romance (Fig. 6.5).

(14) Nervousness

This is a story about the protagonist Nino (the female student who appears at the top left), who is a loner and is often teased and bullied by her

Fig. 6.5 Rin Mikimoto, Kyo no Kira-kun, 42

classmates. On this page, the onomatopoeia ドキン (*dokin*), which denotes the state of emotion (nervousness), appears twice. The onomatopoeia *dokin* first appears across the sheet of stickers Nino is holding and is used again in the panel where she is having a flashback. The *saying* element of *dokin* communicates the nervousness, while the *showing* element of the expression communicates what this nervousness might feel like, via the perceptual resemblance between the sound of the expression and the state of nervousness. On the preceding page, Nino has found a sheet of stickers which she tries to return to the original owner, who, incidentally, is talking with her friend, saying that someone must have stolen the sheet of stickers. This page depicts Nino, the protagonist, having a flashback from her bullied past, while contemplating returning the stickers she has found to the person who is saying they might have been stolen. In both uses of *dokin*, the onomatopoeia is presented in a sharp font, indicating the piercing sense of her nervousness. When it is used for the second time, the onomatopoeia is presented in a much bigger font. This change of font size triggers the search for relevance and guides the readers along the path of least effort until they recover the impression of Nino's emotional state (being increasingly nervous). This suggests that, in addition to the *showing* and *saying* elements of *dokin* as an onomatopoeic word, the visual element of the onomatopoeic presentation has a further showing element that provides direct evidence for communication, illustrating the perceptual resemblance between the typographical features and the state of Nino's emotion, such as increasing nervousness or a piercing sense of nervousness.

Next, in (15), the onomatopoeic expression しゅん (*shun*) is used to denote the feeling of failure. The definition of *shun*, according to the *Weblio Dictionary*, is "in a dispirited manner without hope". The *showing* element of *shun* further communicates how dispirited the subject appears, via the perceptual resemblance between the sound of the expression and the dispirited state (Fig. 6.6).

(15) Sense of failure

This is taken from a monthly manga magazine, *Gekkan Korokoro Comic*. This magazine is typical of weekly or monthly manga publications and contains a number of different manga stories. Its target readers are young boys from primary school to very early teens. This manga series is about a protagonist who causes havoc by his mischievous behaviour, although his heart is in the right place. In this episode, the protagonist tries to help out at lunchtime at school. However, despite his best efforts, he causes calamity rather than being a help. This page depicts the scene where the protagonist is trying to explain what he intended (not) to do. In the middle panel on the right, he declares "I didn't mean to ruin school dinner" when the onomatopoeia *shun* is placed next to his face, written in thin, brush-like strokes, as marked in the image. The use of this expression communicates the dispirited, hopeless manner communicated by the coded element and the particular quality of dispiritedness communicated through the perceptual resemblance between the sound and the

Fig. 6.6 Asumi Yoshino, Yarisugi Itazura Kun, published in Gekkan Korokoro Comic 2018, vol. 4 (April), 50

emotion. In addition, the use of the thin, brush-like stroke of the typeface guides readers to recover the impression that the protagonist is feeling a sense of failure after the havoc he has created. Again, the typographical feature of the onomatopoeia contributes to the communication of an impression, allowing for sharing of the atmosphere in the scene.

Finally, one cannot ignore the onomatopoeia for silence in the discussion of onomatopoeia in manga. *Shi-n* is an onomatopoeia for silence and is often used in manga. Of course, there is no sound of silence that the sound of *shi-n* perceptually resembles. Instead, what it resembles is

Fig. 6.7 Rin Mikimoto, Kyo no Kira-kun p. 27

humans' perception of silence—the sense of continuity (from the long vowel) and whispering sound of /sh/. See example (16), where the onomatopoeia シーン (*shi-n*) is used (Fig. 6.7).

(16) Onomatopoeia for silence

This extract is taken from the same manga series as example (14). In the preceding pages, the protagonist Nino had been told that her classmate Kira, who lives next door, does not have long to live. Also, earlier in the day at school, Kira kisses Nino in front of his girlfriend in order to end their relationship. In this scene, Nino is back from school and is in her room, totally confused about what happened at school, as well as the news on Kira's illness because he did not seem to be that ill, or dying. Towards the end of this page, Nino's pet bird angrily goes next door to wreak vengeance. The onomatopoeia for silence, *shi-n*, is used at the top left, as marked in the image just above Nino's head, while she is sitting on her bed. In addition, there is a speech bubble with dots only, indicating that Nino is not talking. The use of *shi-n* ensures that readers will know that there is absolutely no sound or noise in the room via its coded element, while also *showing* the prolonged nature of the silence. Furthermore, the design of the typeface—the outlined character with no shading—further communicates the impression of emptiness, reinforcing the impression of silence.

As these examples illustrate, onomatopoeia in manga provides evidence for communication at multiple levels. As a linguistic item, whether highly established or not, onomatopoeia involves both *showing* and *saying* to a varying degree, providing both direct and indirect evidence for communication via its perceptual resemblance to the sensory experience in question in the case of manga. The visual element of onomatopoeia in manga provides further direct evidence.

4 Conclusion

This chapter set out to examine how the use of onomatopoeia contributes to creating impressions of scenes in manga, which is the height of Japanese multimodal media. The hybrid representation of visual and verbal elements and narrative in manga (cf. MacWilliams 2008) allows for onomatopoeia to be presented in a highly stylised manner. As the examples in this chapter illustrate, onomatopoeia in manga has a complex layer of representations. To begin with, as discussed throughout this book, onomatopoeia is located on the *showing–saying* continuum (cf. Wharton 2009) and, hence, most onomatopoeia has both *showing* and *saying* elements, providing both direct and indirect (or coded) evidence for communication. In addition, onomatopoeia in manga is often presented in a highly stylised manner as part of the aesthetics, being a prime example of the hybrid representation of the visual and verbal elements that manga is.

The visual element (or the typographical feature) of onomatopoeia gives rise to further effects, by triggering the search for extra effects and attracting the reader's attention. Such visual features guide the reader to follow the path of least effort until their expectation of relevance is satisfied; that is, until they have reached an intended interpretation that justifies the extra processing effort.

The analysis of onomatopoeia in manga, in terms of the relevance-theoretic notion of the *showing–saying* continuum, arguably suggests why onomatopoeia was particularly suited to the visual nature of Japanese culture and the hybrid nature of manga, which involves complex layers of visual and verbal representations. Onomatopoeia's multi- and cross-modal nature allows for the bridging of the senses, which naturally falls into place in visually oriented Japanese culture and media.

Bibliography

Berndt, Jaqueline. 2011. Manga as "Media Geijutsu". *The Locality and Universality of "Media Geijutsu": Beyond "Cool Japan.* ICOMAG 2011.

Cohn, Neil. 2014a. The Architecture of Visual Narrative Comprehension: The Interaction of Narrative Structure and Page Layout in Understanding Comics. *Frontiers in Psychology* 5: 680. https://doi.org/10.3389/fpsyg.2014.00680.

———. 2014b. Visual Narrative Structure. *Cognitive Science* 37 (3): 413–452.

Cooper-chen, Anne M. 2010. *Cartoon Cultures: The Globalization of Japanese Popular Media*. New York: Peter Lang Pub Inc.

Eisner, Will. 2008. *Comics and Sequential Art: Principles and Practices from the Legendary Cartoonist.* 1st ed. New York: W. W. Norton & Company.

Fukuma, Yasuko. 1993. Manga-Ni Miru Giongo/Gitaigo-No Tokuisei-Ni Tuite [On the Peculiarity of Mimetics in Cartoons]. *Kyuusyuu Daigaku Ryuugakusei-Kyooiku-Sentaa Kiyoo [Bulletin of the International Student Center of Kyushu University]* 5: 185–196.

Gravett, Paul. 2004. *Manga: Sixty Years of Japanese Comics.* 1st ed. New York: Harper Design.

Guynes, Sean A. 2014. Four-Color Sound: A Peircean Semiotics of Comic Book Onomatopoeia. *The Public Journal of Semiotics* 6: 58–72.

House, Jill. 2006. Constructing a Context with Intonation. *Journal of Pragmatics* 38 (10): 1542–1558.

Hyuga, Shigeo. 1986. Manga No Giongo/Gitaigo (2) [Onomatopoeia in Manga]. *Nihongogaku* 15 (8): 98–108.
Itoi, Michihiro. 1989. Buntai to Shite Mita "Manga No Kotoba" [Language of Manga as a Style]. *Nihongogaku* 8–9: 66–73.
Kinsella, Sharon. 1999. Pro-Establishment Manga: Pop-Culture and the Balance of Power in Japan. *Media, Culture & Society* 21 (4): 567–572. https://doi.org/10.1177/016344399021004007.
Kitamura, H. 1987. Television Is a Fiction and Viewers Are Observers. *Studies of Broadcasting* 23: 141–153.
Kusamori, Shinichi. 1968. Sutoori Manga No Bunpo Ron. Giseigo Gitaigo No Ronri [Grammar of Story Manga: Logic of Onomatopoeia]. *Com.*
MacWilliams, Mark W. 2008. *Japanese Visual Culture: Explorations in the World of Manga and Anime*. London: Routledge. https://www.goodreads.com/work/best_book/3148567-japanese-visual-culture-explorations-in-the-world-of-manga-and-anime.
McCloud, Scott. 1993. *Understanding Comics: The Invisible Art*. Princeton, NJ: Kitchen Sink Press.
———. 2006. *Making Comics: Storytelling Secrets of Comics, Manga and Graphic Novels*. New York: Harper.
Morooka, Tomonori. 2010. The Onomatopoeias in Manga. *Studies in Literature and Culture* 47: 15–24.
Natsume, Fusanosuke. 1997. *Manga No Bunpo: Komikkugaku No Mikata [Grammar of Manga: Perspectives on Comic Studies]*. Tokyo: Asahi Shimbunsha.
———. 2013. Manga Ni Okeru Onomatope [Onomatopoeia in Manga]. In *Onomatope Kenkyu No Shatei: Chikazuku Oto to Imi*, ed. Kazuko Shinohara and Ryoko Uno, 217–241. Tokyo: Hitsuji Shobo.
O'Hagan, Minako. 2010. Japanese TV Entertainment: Framing Humour with Open Caption Telop. In *Translation, Humour and the Media: Translation and Humour*, ed. Delia Chiaro, vol. 2, 70–88. London: Continuum.
Petersen, Robert S. 2009. The Acoustics of Manga. In *A Comics Studies Reader*, ed. Jeet Heer and Kent Worcester, 163–172. Jackson, MS: University Press of Mississippi. http://www.jstor.org/stable/j.ctt2tvd9s.22.
Pratha, Nimish, Natalie Avunjian, and Neil Cohn. 2016. Pow, Punch, Pika, and Chu: The Structure of Sound Effects in Genres of American Comics and Japanese Manga. *Multimodal Communication* 5 (October): 93–109. https://doi.org/10.1515/mc-2016-0017.
Rohan, Olivia, Ryoko Sasamoto, and Rebecca Jackson. 2018. Argumentation, Relevance Theory and Persuasion: An Analysis of Onomatopoeia in Japanese

Publications Using Manga Stylistics. *International Review of Pragmatics* 10 (2): 219–242.
Sasamoto, Jun. 2000. Manga Hyogen Ni Okeru "Gion" No Hataraki [On the Function of "Gion" in Manga; Focusing on Its Role Word]. *Vi*.
Sasamoto, Ryoko. 2014. Impact Caption as a Highlighting Device: Attempts at Viewer Manipulation on TV. *Discourse, Context & Media* 6: 1–10.
Sasamoto, Ryoko, and Rebecca Jackson. 2016. Onomatopoeia—Showing-Word or Saying-Word? Relevance Theory, Lexis, and the Communication of Impressions. *Lingua* 175: 36–53.
Sasamoto, Ryoko, Minako O'Hagan, and Stephen Doherty. 2017. Telop, Affect, and Media Design: A Multimodal Analysis of Japanese TV Programs. *Television & New Media* 18 (5): 427–440. https://doi.org/10.1177/1527476416677099.
Schodt, Frederik L. 1983. *Manga! Manga!* New York: Kodansha America https://www.goodreads.com/work/best_book/948173-manga-manga-the-world-of-Japanese-comics.
———. 1996. *Dreamland Japan: Writings on Modern Manga*. Berkeley, CA: Stone Bridge Press.
Schwartz, Adam, and Eliane Rubinstein-Ávila. 2006. Understanding the Manga Hype: Uncovering the Multimodality of Comic-Book Literacies. *Journal of Adolescent & Adult Literacy* 50 (1): 40–49. https://doi.org/10.1598/JAAL.50.1.5.
Scott, Kate. 2017. Prosody, Procedures and Pragmatics. In *Semantics and Pragmatics: Drawing a Line*, ed. Ilse Depraetere and Raphael Salkie, 323–341. Berlin: Springer International. https://doi.org/10.1007/978-3-319-32247-6.
Scott, Kate, and Rebecca Jackson. Forthcoming. When Everything Stands Out, Nothing Does: Typography, Expectations and Procedures. In *Relevance Theory and Figuration*, ed. Piskorska Agnieszka. Amsterdam: John Benjamins Publishing.
Shiota, Eiko. 2003. Relevance Theory and Understanding Telops. *The Graduate School Review of the English Language and Literature* 31 (March): 63–91.
Shitara, Kaoru. 2011. The Change of Telops on NHK Variety Shows. *Bulletin of Mukogawa Women's University, Humanities and Social Science* 59: 1–9.
Sontag, Susan. 1999. The Image World. In *Visual Culture: The Reader*, ed. Jessica Evans and Stuart Hall, 80–94. London and Thousand Oaks: Sage Publications in Association with the Open University.

Sperber, Dan, and Deirdre Wilson. 1986/1995. *Relevance: Communication and Cognition*, 2nd ed. Oxford: Blackwell.

———. 2015. Beyond Speaker's Meaning. *Croatian Journal of Philosophy* 15 (2): 117–149.

Tajima, Kaori. 2006. Onomatope (Giongo Gitaigo) Nit Suite. [On Onomatopoeia (Mimetics)]. *Kansai Gaikokugo Daigaku Ryuugakusei-Bekka Nihongo-Kyooiku Ronsyuu [Studies in Japanese Language Education at the International Center of Kansai Gaidai University]* 16: 193–205.

Wharton, Tim. 2009. *Pragmatics and Non-Verbal Communication*. Cambridge: Cambridge University Press. https://doi.org/10.1017/CBO9780511635649.

Yus, Francisco. 2009. Visual Metaphor Versus Verbal Metaphor: A Unified Account. In *Multimodal Metaphor*, ed. C. Forceville and E. Uriós-Aparisi, 145–172. Berlín: Mouton de Gruyter.

7

Onomatopoeia and Translation: A Corpus Approach

The fact that onomatopoeia is so intricately linked to our sensory experience indicates that it might be hard to capture in words, which might cause some concerns for translators. Indeed, it is often argued that the translation of onomatopoeia poses particular difficulties for translators. However, previous studies on the translation of onomatopoeia seem to be focused on literary translation, and, as such, most of the analyses are limited to the examination of the uses of onomatopoeia by certain authors, comparing original and translated texts, even when they are taking a corpus-based approach (e.g. Casas-Tost 2014). As a result, there is little research that has examined the more general translation of onomatopoeia. Similarly, most studies on the translation of onomatopoeia are based on a descriptive taxonomic approach, where the focus is to classify likely translation strategies or the likely word class with which the onomatopoeia is translated.[1] Therefore, there is a paucity of explanatory research into why and how onomatopoeia is omitted from translation or why the translation of onomatopoeia is difficult. Furthermore, it is not empirically shown that onomatopoeia is indeed a challenge for translators.

[1] Sato (2017) takes a different approach to onomatopoeia translation and analyses onomatopoeia and translation from the perspective of translanguaging.

R. Sasamoto, *Onomatopoeia and Relevance*, Palgrave Studies in Sound,
https://doi.org/10.1007/978-3-030-26318-8_7

Against this backdrop, this chapter is an attempt to investigate how and why onomatopoeia is translated (or not) in user-generated data. A parallel corpus was built using user-generated data from the popular Japanese cooking website Cookpad Inc. and its translation into English. Cookpad is one of the biggest recipe sites in Japan and hosts over 30,000 recipes. The website is bigger than other global recipe sites such as Yummly. However, before presenting my own analysis, I will first review the work of scholars who have taken corpus-based approaches to the analysis of onomatopoeia with a particular focus on translation.

1 A Challenge for Translation? Previous Studies on Onomatopoeia Translation

It has often been claimed, sometimes anecdotally, that the translation of onomatopoeia can be challenging. As Bartashova and Sichinskiy (2014) argue, problems are considered to arise when the target language does not appear to have any onomatopoeic lexical resources allowing for a one-to-one translation of the onomatopoeic item in the source text. Similarly, scholars such as Edström (1989), Hayase (1978), and Flyxe (2002) report that it is not uncommon for translators to leave onomatopoeia untranslated. Kubo (1995) even reports that 78% of onomatopoeic expressions in the works of Kenji Miyazawa are left untranslated. Indeed, I showed such a case in Chap. 3, with the example of *Matasaburo of the Wind*:

(1) [Describing the strong wind that is storming through the village]
Doddodo dodō do, dodoōdo, dodō aoi kurumi mo fukitobase
MIM blue chestnuts also blow-away
Suppai Karin mo fukitobase *Doddodo dodō do, dodoōdo, dodō*
sour quince also blow-away MIM
"Doddodo dodō do, dodoōdo, dodō; Blow away the green chestnuts too;

Blow away the sour quince too; Doddodo dodō do, dodoōdo, dodō…."
(Miyazawa Kenji, *Kaze no Matasaburo* [Matasaburo of the Wind], 1934, 1.
English translation by Strong and Colligan-Taylor, 2002)

This example illustrates how the onomatopoeic expression *Doddodo dodō do, dodoōdo, dodō* is left untranslated and is kept in the English version. This is a case of the transliteration of onomatopoeia. In other cases, such as (2), onomatopoeia can be left out of the translation completely:

(2) "Sorekara hōkiboshiga *gī-gī-fūgī-gī-fūQte* ittekita ne."
then, comet-NOM gī-gī-fūgī-gī-fū-QUO say-and came CONF
"And then a comet zooms by."
(Miyazawa Kenji, *Ginga Tetsudo no Yoru* [Night on Milky Way Railroad]'
English translation by Shelley Marshall (2014), cited in Sato 2017, my italics)

In (2) the onomatopoeia *gī-gī-fūgī-gī-fū*, which describes the manner in which the comet shoots across the sky, is omitted from the target text. This is a case of omission.[2]

Scholars such as Bartashova and Sichinskiy (2014), Flyxe (2002), and Inose (2008) are concerned with the challenges of translating onomatopoeic words in the source text into adequate renderings in the target text. Bartashova and Sichinskiy (2014, 223–226) argue that the difficulty in translation is a result of a style/register difference or the lack of equivalent phonetic symbolism. As such, the focus of research into the translation of onomatopoeia is often on identifying and describing a taxonomy of types of translation strategies. This is done, naturally, in an attempt to provide systematic and reliable guidance to translators about what to do in order to create an optimal translation in particular linguistic circumstances.

In an attempt to establish a taxonomy of translation strategies or techniques, most scholars take corpus-based approaches. Flyxe (2002) conducted an empirical analysis of the translation of onomatopoeia in

[2] As I will discuss later in this chapter, omission is considered one of the appropriate translation strategies.

Japanese literary texts into Swedish and analysed 257 expressions in relation to the lexicality of onomatopoeic expressions and translation methods. He used Yoshimoto Banana's *Kitchen*, Endo Shusaku's *Sukyandaru*, Miyazawa Kenji's *Nametoko Yama no Kuma*, *Shikaodori no Hajimari*, and *Yodaka no Hoshi*, where he analysed the translation of onomatopoeia in terms of the degree of lexicalisation discussed in Chap. 4, first proposed by Tamori and Schorup (1999). According to Flyxe (2002, 56), lexicality is "the degree to which a word is part of the established vocabulary, and mimeticity, or the degree to which a word is a direct imitation of a sound, state or condition". As Flyxe (2002) explains, lexicality and mimeticity are located at opposite ends of a continuum. He further notes (2002, 56) that "the more an onomatopoeia is part of the established vocabulary, the less mimetic it is" and that an "increasing degree of mimeticity also points at a description of something concrete rather than abstract".

Flyxe (2002) observes that there are six methods involved in translating onomatopoeia and mimetics: (a) adjectives, (b) adverbs, (c) verbs, (d) explicative paraphrases, (e) onomatopoeia and mimetics, and (f) omissions. The summary of Flyxe's (2002) findings is given in Table 7.1.

According to Flexe (2002), *giongo* (an auditory-based onomatopoeia) and *giongo+to* (an auditory-based onomatopoeia with an obligatory quotative particle *to*) are considered to be located at the lower end of lexicalisation, while CVQ/NCV-*ri*, *gijogo* (psychomimes) and those with the *suru*-verb are considered highly lexicalised. Flyxe (200) argues that highly lexicalised onomatopoeia are more difficult to translate. Therefore, *gijogo* (psychomimes) and *gitaigo* (non-auditory-based onomatopoeia), which

Table 7.1 Summary of Swedish translations (*N177*)

	CVQ/ NCV-ri	*Ijōgo* (psychomimes)	*Suru*- verb	*Giongo* (all) (phonomimes)	*Giongo* +*to*	*Giongo*
Adjectives	28	16	21	0	0	0
Adverbs	30	4	1	0	0	0
Verbs	4	3	3	16	7	9
Paraphrases	17	13	7	4	0	4
Onomatopoeia	0	0	0	2	2	0
Omissions	4	0	0	4	2	2
Totals	83	36	32	26	11	15

Source: Based on Flyxe (2002, 70)

are highly lexicalised, are abstract and, as such, they are often translated using a paraphrase, since finding equivalent terms in the target language is difficult. He also suggests that the issue with register leads to challenges for translating onomatopoeia, as onomatopoeia in Swedish is too informal to use in literary translation, given that "pure onomatopoeia in Swedish has a childish, sometimes vulgar, nuance and might therefore be avoided" (Flyxe 2002, 71). Another possible cause of the difficulties in translating onomatopoeia is the differences in the phonetic systems of Japanese and Swedish. Flyxe (2002) suggests that the Japanese sound structure system is far more complex than that of Swedish and, as such, Japanese sound symbolism can express rich varieties of nuances.

While Flyxe's analysis in terms of the degree of lexicality sheds some light on the relationship between lexicality and translation methods, it is still not entirely clear what it means to be "difficult to translate". Just because something is translated using a paraphrase, does it mean that the translation was difficult? Furthermore, his data contain more "highly lexicalised onomatopoeia" than the expressions that are located at the lower end of lexicalisation. The examination of a wider range of examples would be necessary to address this issue.

Minashima (2004) analyses 332 onomatopoeic expressions used in Banana Yoshimoto's *Kitchen* and how these are translated. His findings are summarised in Table 7.2.

According to Minashima (2004), onomatopoeia is translated using adverbs, adjectives, and nouns in most cases, and translation using ono-

Table 7.2 Summary of onomatopoeia translations

	Animate phonomime	Inanimate phonomime	Phenomime	Total (%)
Verb	3	13	74	90 (27.1%)
Adverb	1	0	59	60 (18.1%)
Adjective	0	1	50	51 (15.4%)
Noun	0	3	31	34 (10.2%)
Onomatopoeia	1	19	4	24 (7.2 %)
Other	0	0	19	19 (5.7 %)
No translation	0	5	49	54 (16.3%)
Total	5	41	286	332 (100%)

Source: Based on Minashima (2004, 112, my translation)

matopoeia is uncommon. However, it is not unusual for onomatopoeia to be left untranslated or omitted. Indeed, his findings show that 16.3% of all onomatopoeia are left untranslated.[3] Such findings are similar to those reported by other scholars.[4]

Inose (2008) also conducted a corpus-based analysis of onomatopoeia,[5] using a corpus of onomatopoeic expressions from Haruki Murakami's *Sputnik no koibito* (1999) and comparing them with English and Spanish translations. Her Japanese corpus contains 295 onomatopoeic expressions, and she identifies nine translation methods:

(1) Onomatopoeia used as an adverb or translated using adverbs (77.3% in Japanese (JP), 16.3% in English (EN), 21% in Spanish (SP))
(2) Onomatopoeia used as adjectives or translation using adjectives (4.4% in JP, 15.3% in EN, 18% in SP)
(3) Onomatopoeia used as verbs or translation using verbs (16.3% in JP, 16.6% in EN, 10.8% in SP)
(4) Onomatopoeia used as nouns or translation using nouns (0.3% in JP, 4.4% in EN, 3.1% in SP)
(5) Translation using explicative paraphrases (20% in EN, 12.5% in SP)
(6) Translation using idioms (not many in either language)
(7) Translation using onomatopoeic expressions (16 cases in EN, 8 in SP)
(8) Translation using two adjectives, or combination of adverbs, adjectives, or verbs (7 instances in EN, 5 instances in SP)
(9) No translation (omission or a complete change of the phrase) (19.3% in EN, 19.3% in SP; 7.1% of omission cases in both languages)

As Inose's findings show, translating onomatopoeia, as with any other translation, does not assume equivalence in grammatical categories. For

[3] Minashima (2004) speculates that the relatively high rate of omission might be related to translators' unfamiliarity with onomatopoeia.

[4] Minashima's (2004) findings also show that it is very rare to translate non-auditory-based onomatopoeia (or mimetics in the sound-symbolism-based approach) using onomatopoeia; 46.3% of onomatopoeic expressions for inanimate sound are translated using equivalent onomatopoeia. This finding is in line with Takeuchi's (1998, cited in Flyxe 2002) study on the translation of Japanese onomatopoeia into French, which shows that more complex translation methods were used for non-auditory-based onomatopoeia. This, perhaps, is related to the determinateness and accessibility of the source experience, which lead to the existence of similar expressions in both source and target languages.

[5] As with most scholars, Inose (2008) distinguishes between onomatopoeia and mimetics.

example, while most (77.3%) onomatopoeic expressions in the original Japanese function as adverbs, this is not the case for the translation, as only 16.3% in English and 21% in Spanish are translated using adverbs. In addition, almost half of onomatopoeia in the original are translated using methods other than adverbs, adjectives, verbs, or nouns (47.5% in English, 47.1% in Spanish), when only 1.7% of the onomatopoeia in the original Japanese function in any way other than adverbs, adjectives, verbs, or nouns. According to Inose (2008, 114), translation using adverbs, adjectives, verbs, or nouns is more "straightforward" in cases where translators can find the "equivalent" in the target language. She seems to consider that the fact that almost half of the original onomatopoeia are translated using non-straightforward methods is an indication of the challenges of translating onomatopoeia. She further argues that in cases where the equivalent in the target language is not readily available, translators opt for "the use of 'more original' translation methods rather than assigning an equivalent word in the target language".

Inose (2008) is the only researcher to further examine omission cases. She argues that omission is not necessarily a simple solution and compares the number of omissions in each translation (Table 7.3).

Inose argues that most instances of omission were found in uses of onomatopoeic expressions as a modifier, rather than in those carrying the essential information. She argues that these are "cases where those expressions do not add new information to the phrase but elaborate, or even repeat the information already given by other words (verb, for example)" (Inose 2008, 114). This may suggest that translators might have seen it as appropriate to omit the onomatopoeic expression, despite the loss of subtle nuances. She therefore concludes that omission is not the most desirable translation method and that, "when there is no equivalent word

Table 7.3 Summary of omissions by Inose

	English		Spanish		Both	
	No.	%	No.	%	No.	%
Onomatopoeia	4 (of 28)	14.3	8 (of 28)	28.6	3 (of 28)	10.7
Mimetic words	53 (of 267)	19.9	49 (of 267)	18.4	18 (of 267)	6.7
Total	57 (of 295)	19.3	57 (of 295)	19.3	21 (of 295)	7.1

Source: Based on Inose (2008, 112)

in the target language, the translators should consider using other resources, such as explicative paraphrases or combination of various words" (Inose 2008, 115).

So far, I have reviewed studies that are concerned with the translation of onomatopoeia from Japanese. In contrast, Casas-Tost (2014) presents a corpus-based analysis of onomatopoeia translation from Chinese to Spanish. Her corpus consists of seven contemporary Chinese novels and their Spanish translations, and contains 490 onomatopoeia in Chinese.

Using Toury (2004) and Molina's (2006) adequacy-acceptability conceptual framework, Casas-Tost (2014, 44) identifies eight translation methods of onomatopoeia: amplification (e.g. explanatory paraphrasing); discursive creations (e.g. an ephemeral equivalent); description (e.g. replacing the term with a description of its form/function); established equivalent; generalisation (e.g. the use of a more general/neutral term); borrowing (incorporating the expression from the source language); reduction (e.g. suppressing information partially or completely); and substitution (i.e. translating using a form other than onomatopoeia). According to Casas-Tost (2014), the three most common translation methods are substitution (50.6%), omission (32.6%), and translating a source-text onomatopoeia by using a target-language onomatopoeia (16.7%).

Despite the fact that Casas-Tost (2014) is concerned with a different language pair (Chinese–Spanish) from other studies, the result gives very similar findings to others which focus on the Japanese–English pair. That is, onomatopoeia is rarely translated using equivalent onomatopoeia. Instead, it is invariably translated using other forms or it is left untranslated.

So far, I have reviewed studies on the translation of onomatopoeia, most of which take a taxonomic approach. While such studies offer a rich description of how onomatopoeia can be translated, the topic of onomatopoeia translation is evidently still under-researched and does not account for cases where onomatopoeia is left out of translation, whether omitted or transliterated. Furthermore, studies on the translation of onomatopoeia have so far focused largely on literary texts and the sample size examined in previous studies is relatively small. This means that the range of data examined is limited to literary texts written by professional authors or translated by professional translators. That is, not only are these studies limited in the size of the sample, they are also limited to a very small number of individuals who are acclaimed language users. It is therefore not clear how representative they are as language users. In addition, all literary works that are examined in previous studies are translated by a native speaker of the target language, with the exception of one Chinese title (*Half of Man Is Woman*). While there is no study on the directionality of translation and its influences on the translation of onomatopoeia, the culturally bound nature of onomatopoeia would suggest that individuals' experience with onomatopoeia in their native language might affect the manner in which they translate onomatopoeia.

Furthermore, all the works on the translation of onomatopoeia are concerned with how onomatopoeia in the source texts are translated, and they never discuss how items that are not onomatopoeia in the source language might be translated using onomatopoeia. Such a discussion would be extremely useful to understand how the communication of impressions is effected in different ways across languages. However, little empirically supported analysis has been conducted so far.

Another issue is that these studies seem to equate the group of linguistic items that come under the name of onomatopoeia as a part of speech, except for Inose (2008). Such an assumption is evident from the way they analyse the translation of onomatopoeia being carried out—creating a taxonomy of how *onomatopoeia* is translated, be it an adverb, verb, or another onomatopoeia, and regardless of the role of each onomatopoeic expression in the source language. Such an analysis suggests that authors assume onomatopoeia as one grammatical category or equivalent when it is, in fact, a characteristic of words and can appear across different grammatical functions. Moreover, the analysis of onomatopoeia translation in terms of grammatical categories assumes equivalence in grammatical categories between two languages. It is not clear whether finding a grammatical category of onomatopoeia translation would be useful for a distant language pair such as Japanese and English, which have entirely different linguistic structures.

The most fundamental issue with these studies is the underlying assumption that onomatopoeia is a challenge for translators. This idea might be advantageous to researchers working in pragmatics and in pragmatic stylistics, as the so-called challenges of translating onomatopoeia might be indicative of the assumption that the phenomenon is linked with the communication of vague, weak, impressionistic, and/or non-propositional effects. This suggests the need for further investigation of onomatopoeia and its interpretation in more detail. However, the real question is: does the absence of one-to-one equivalence in terms of lexical items necessarily suggest a challenge for translators, and hence a loss of meaning? Some scholars seem to consider the lack of onomatopoeia in the target language, or the difference in the phonetic system, as the cause of the challenge (cf. Bartashova and Sichinskiy 2014; Flyxe 2002). Similarly, others seem to consider that the lack of "equivalence" would pose a challenge for the translator, as explicitly argued by Inose (2008). She has shown how more than half of onomatopoeia uses in her corpus were translated using a non-straightforward strategy, where translators cannot use an "equivalent" lexicon in the target language in categories such as adverbs, adjectives, verbs, or nouns, and have to resort to the translation of onomatopoeia using more original translation methods. Of course, the lack of formal equivalence would mean non-straightforward translation, to which Inose (2008) alludes. However, does this necessarily impose challenges to successful translation? After all, effective/successful translation can often be achieved via indirect equivalence.

To shed light on issues surrounding the translation of onomatopoeia, and, most importantly, to examine where onomatopoeia does (or does not) post a challenge to translators, the remainder of this chapter will focus on a corpus-based analysis using recipe data from Cookpad Inc. I conducted a corpus analysis using the parallel corpus of approximately 16,000 recipes in Japanese and their translation into English. The corpus was provided by Cookpad Inc. This is a relatively small corpus (3,371,255 words in Japanese, 3,981,680 words in English), as most parallel corpora are built for machine learning. However, to the best of my knowledge, there is no analysis of onomatopoeia translation using a corpus of this size. As discussed, most corpora used in previous studies are much smaller and consist only of extracts from a few literary works. In addition, recipes in Cookpad are uploaded by the users. As Harashima et al. (2016) explain, Cookpad is a popular website for searching recipes, with more than 3 million recipes written in Japanese as of May 2019. Users can upload their recipes and also comment on recipes uploaded by other users. In addition to recipes uploaded by users, Cookpad also offers recipes written by professional chefs and nutritionists, responding to the specific needs of users. According to Toyo Keizai Online (26 February 2016), more than 57 million users use Cookpad each month. Cookpad also offers its service in 22 different languages, including Arabic, English, French, Korean, Polish, and Vietnamese. Cookpad, especially its original Japanese site, has a very strong social networking aspect, where users comment on each other's recipes and follow certain users. The use of this corpus addresses some of the issues already discussed, especially the size of the corpus and the real-life nature of language use.

2 Corpus-Based Approach to Onomatopoeia

2.1 Methodology

SketchEngine[6] was used as a corpus tool. First, the word list was generated to identify all of the onomatopoeia that appear in the Cookpad corpus.

There were 202 lema in total (Table 7.4), each of which had a number of variants (e.g. torori, Torotoro, torottoro, toroori).

The onomatopoeic expressions listed in Table 7.4 were categorised based on Flyxe's (2002) degrees of lexicalisation—highly lexicalised onomatopoeia (Group 1), middle-range onomatopoeia (Group 2), and less lexicalised onomatopoeia (Group 3)—since his work indicates that there is a relationship between the level of lexicalisation and translation strategies. For the purpose of this study, five expressions from each category were chosen for analysis, based on frequency (Table 7.5). It should be noted that there was some overlap and these categories are on a continuum, rather than there being a clear-cut distinction between them.

Initial searches on all onomatopoeic expressions (as listed on SketchEngine) produced a huge concordance (39,668 onomatopoeia used in total). In this study, the detailed analyses were conducted only on random samples. The sample size was calculated using the Sample Size

[6] www.sketchengine.eu

Table 7.4 A full list of onomatopoeia found in Cookpad

AQ[a]sari	KoQkuri	Jinwari	Pakipaki	Betobeto
Ahaha	Kotsukotsu	Sū	Bakubaku	Perapera
UQkari	KoQsori	Sukasuka	Pakupaku	Perori
UQsura	KoQteri	SuQ	Basabasa	Hokahoka
UQtori	Kotokoto	ZuQ	Pasapasa	Pokapoka
Gā	Konekone	SuQkari	Batan	Hokuhoku
Kachikochi	Korikori	suQkiri	Pachipachi	Bokoboko
gaQ	Korokoro	zuQshiri	paQ	Bosoboso
Gatsugatsu	Gorogoro	suQpori	paQ	Posoposo
gaQtsuri	Kongari	Surusuru	haQkiri	Potapota
karaQto	Kon-mori	Sō	Hafuhafu	hoQ
Karari	Sakusaku	Sorosoro	Barabara	poQ
Garigari	Zakuzaku	taQpuri	Parapara	hoQkuri
Karikari	sasaQ	Damadama	Baribari	hoQkori
Gangan	zaQkuri	Taratara	Paripari	boQteri
Gizagiza	saQsa	Daradara	Pikapika	poQteri
Kichin	saQ	chachaQ	Hitahita	Potsupotsu
giQshiri	zaQ	Chanto	Pitapita	Potopoto
kiQchiri	saQpari	Choichoi	biQkuri	Poripori
kiQchiri	Sarasara	Chokichoki	biQshiri	Poripori
Gitogito	Zarazara	Chonchon	piQchiri	Horohoro
Gyūgyū	Shikoshiko	Chirichiri	Piripiri	Boroboro
Kira	shiQkari	Chin	Fukafuka	Poroporo
Girigiri	shiQkuri	Tsubutsubu	Bukubuku	Howahowa
kiriQ	jiQkuri	Tsuyatsuya	Pukupuku	Honnori
Guigui	jiQ	tsuruQ	Busubusu	Ponpon
Kutakuta	shiQtori	Dōn	Purupusu	Honwaka
Guchafucha	Shimijimi	Dokidoki	Puchipuchi	Honwari
guQ	Ja	doQsari	fuQkura	maQtari
kuQkiri	Shakishaki	doQshiri	puQkuri	muQchiri
Kutsukutsu	Shakushaku	doQpuri	Futsufutsu	Munyu
Gutsugutsu	Shabushabu	dobaQ	Butsubutsu	Mufu
Guragura	Sharishari	Torori	Putsuputsu	Mokomoko
Guriguri	Jarijari	Dorori	Furifuri	Mochimochi
Kurukuru	Ju	Tonton	Furufuru	moQtari
Guruguru	Shuwa	Dondon	Purupuru	Mofumofu
Kukruri	Juwa	Naminami	Fuwafuwa	yuQkuri
Gururi	Shunshun	nikoniko	Betabeta	yuQtari
Kurun	Shiwashiwa	neQtori	Bechabecha	Yuruyuru
Gun	Jiwajiwa	Nebaneba	beQtori	Waiwai
Gokugoku	shinnari			

[a]The first part of geminate consonants is represented by /Q/.

Table 7.5 A list of onomatopoeia examined in this study

Lema	Frequency (*n*)	Random sample (*n*) (CI and CL based)	Groups
ShiQkari	5018	536	1
TaQpuri	3124	504	1
ShiQtori	1669	442	1
SaQpari	939	366	1
Shinnari	927	365	1
Torotoro	3120	503	2
Fuwafuwa	3024	501	2
Sakusaku	1994	462	2
SaQ	1177	398	2
Karikari	1000	375	2
Pachipachi	73	65	3
Gā	48	45	3
Puchipuchi	39	37	3
Dōn	23	23	3
Gangan	17	17	3

CI confidence interval; *CL* confidence level

Calculator from Survey System.[7] The confidence interval (CI) was set at 4, and the confidence level (CL) was set at 95%. Unlike previous studies, which focused on classifications of translation strategies or grammatical class, this study analysed how onomatopoeia are translated at a lexical level. Once the target-text expression of onomatopoeia in the source text was identified, it was analysed in terms of (1) the total number of expressions used to translate each onomatopoeic expression; (2) the number of expressions used more than five times (a); (3) the frequency of (a); and (4) the percentage of (a) in random samples. Having identified the target-text expressions and their occurrence, I analysed cases of omission, as the high frequency of omission in previous studies seemed to result in the preconception that onomatopoeia is a challenge for translators. I examined cases of omission in terms of contextual factors that appear to have influenced the translation strategy. I will return to the omission cases later in the chapter, but first, in the next section, expressions that were used to translate onomatopoeia in more than five instances are discussed.

[7] www.surveysystems.com/sscals.html

2.2 Onomatopoeia in the Cookpad Parallel Corpus

2.2.1 Group 1 Expressions: Highly Lexicalised Onomatopoeia

ShiQkari

ShiQkari is a well-established (highly lexicalised) onomatopoeic expression, which can be used to express the strength of the seasoning or the flavour of the dish, the degree of completion of the cooking process, or the action of cooking. There was a total of 5018 instances. For the analysis, 537 random samples were examined. Two examples were discarded due to an error in concordance. In total, 75 expressions were used to translate onomatopoeia in the source text. However, 343 cases (62%) were translated using the 11 expressions in Table 7.6.

TaQpuri

TaQpuri is an expression used to express a large volume and is translated mostly as a "generous amount, filled, full of, lots of, plenty of", and so on. In this corpus, *taQpuri* was used a total of 3124 cases, and a random sample of 504 cases was examined in detail. Two cases were discarded later for errors in concordance. There were 59 different target-text expressions used to translate onomatopoeia in the source text, but more

Table 7.6 Frequently used target-text expressions for *shiQkari*

Frequency (*n*)	Target-text expression
146	well
56	thoroughly
38	tightly
32	firm
25	completely
11	proper
10	strong
7	securely
7	stiff
6	make sure
5	stiff

Table 7.7 Frequently used target-text expressions for *taQpuri*

Frequency (*n*)	Target-text expression
141	a lot of
106	plenty of
52	a/the generous N of
31	full of
22	packed
12	fill
8	a large N of
8	rich with

Table 7.8 Frequently used target-text expressions for *shiQtori*

Frequency (*n*)	Target-text expression
370	moisture
10	soft
5	tender
5	rich

than 380 cases (over 75%) were translated using the eight phrases (and their variants) in Table 7.7.

ShiQtori

ShiQtori is an onomatopoeic expression that is often used to describe a slightly damp texture. In this corpus, *shiQtori* appears 1669 times and 442 random samples were analysed. One sample had to be discarded due to an error in concordance. In total, there are 24 expressions (and their variants) used to translate *shiQtori*. However, the overwhelming majority (390 cases, 88.4%) were translated using the four expressions (and their variants) shown in Table 7.8.

Furthermore, most uses of *shiQtori* (370 times, 83.52%) were translated using a single English expression, *moist* (and its variants such as *moisture*, *moistness*).

SaQpari

SaQpari is an expression often used to describe the way a food or drink does not have any long-lasting flavour, taste, or texture. In this corpus, it

Table 7.9 Frequently used target-text expressions for *saQpari*

Frequency (*n*)	Target-text expression
175	refreshing
74	light
11	light and refreshing
6	simple

Table 7.10 Frequently used target-text expressions for *shinnari*

Frequency (*n*)	Target-text expression
151	wilt
90	soft
70	tender
16	soft and wilted
10	translucent
5	limp

appears 939 times and 366 random samples were examined in detail. There are 29 translation candidates, but the majority (320 cases, 87.4%) were translated using the four phrases (and their variants) listed in Table 7.9.

Shinnari

According to the *Japanese Onomatopoeia Dictionary* (Ono 2007, 196), *shinnari* is often used to describe something that is soft yet has kept its elasticity. It appears 927 times in this corpus and 365 random samples were examined. Later, four samples had to be discarded due to a concordance error.

The translation of this onomatopoeia seems straightforward. While 14 expressions were used to translate *shinnari* in this sample, 342 (93.69%) were translated using the six phraes (and their variants) given in Table 7.10.

Table 7.11 Frequently used target-text expressions for *torotoro*

Frequency (*n*)	Target-text expression
248	thick
59	creamy
46	melt
36	cheese
19	soft
10	creamy and soft
7	juicy
6	thick and creamy
5	tender

2.2.2 Group 2 Expressions: Middle Range

Torotoro

Torotoro is an onomatopoeic expression for texture. In this corpus, *torotoro* and its variants were used 3120 times. A sample of 503 cases were examined and five were discarded, due to errors in concordance. In total, 39 different expressions were used, but over 436 (86.67%) cases were translated using the nine phrases (and their variants) given in Table 7.11.

Fuwafuwa

The onomatopoeic expression *fuwafuwa* is used to describe something light and airy, often concerning bread or cake. While it largely describes food in this corpus, there are cases where the onomatopoeia is used to describe an action. In this corpus, *torotoro* and its variants were used 3024 times. A sample size of 501 cases were examined. In total, 30 different expressions were used, but 447 (89.22%) cases were translated using the seven phrases (and their variants) outlined in Table 7.12. There was a relatively small number of omissions (25 cases, 4.9%).

Table 7.12 Frequently used target-text expressions for *fuwafuwa*

Frequency (*n*)	Target-text expression
313	fluffy
34	light
33	loosely
30	soft
24	fluffy and light
8	airy and light
5	tender

Table 7.13 Frequently used target-text expressions for *sakuQ*

Frequency (*n*)	Target-text expression
174	crisp
48	gently
42	crunch
42	lightly
9	thoroughly
8	flaky
7	well
6	roughly
5	crisp and light

SakuQ

sakuQ is frequently used to describe the crispiness of an object or the swiftness of an action. Its variants include *sakusaku* or *saQkuri*. In this corpus, the expression appears 1994 times and 462 random samples were analysed. One case was discarded due to a concordance error. As with other expressions I have presented so far, many phrases (26 in total) were used to translate this in the source text. However, the majority (341 cases, 73.8%) were translated using the nine expressions shown in Table 7.13. Considering that the last five expressions appear only a limited number of times, *crisp*, *gently*, *crunch*, and *lightly* are the most likely translation candidates.

SaQ

SaQ is almost always used with *to* and is an onomatopoeia to describe swiftness (Ono 2007). In this corpus it appeared 1177 times, and 398

Table 7.14 Frequently used target-text expressions for *saQ*

Frequency (*n*)	Target-text expression
137	quick
49	briefly
19	lightly
10	just
7	briskly

Table 7.15 Frequently used target-text expressions for *karikari*

Frequency (*n*)	Target-text expression
307	crisp
43	crunchy

random samples were analysed. Two cases were discarded due to a concordance error and duplication. While 18 expressions were used to translate this, over half of cases (222 cases, 55.77%) were translated using the five expressions and their variants listed in Table 7.14.

Karikari

Karikari is an expression to denote the state of a dry object. It appears 1000 times in this corpus and 375 random samples were examined further.

Unlike other onomatopoeia which have a wide range of translations, *karikari* has a limited range of English expressions to translate with, and an overwhelming 354 cases (94.4%) are translated using *crisp*, *crunchy*, and its variants (including the combination *crispy crunchy*), as shown in Table 7.15.

2.2.3 Group 3 Expressions: Less Lexicalised Onomatopoeia

Finally, in this section I will examine a range of expressions that are categorised as Group 3, namely *giongo* (phonomimes). Compared with the onomatopoeia in Groups 1 and 2, these are used much less in this corpus, with the most frequent phonomime *pachipachi* being used only 73 times. Still, for consistency of methodology, the same random samples were examined with the CI set at 4 and the CL at 95%.

Pachipachi

Pachipachi is used to represent a sharp, repetitive noise such as fire, clapping, or popcorn. In this corpus there are 73 occurrences in total, and 65 random samples were examined in detail.

It is worth noting that there are no cases of omission. The 12 expressions in Table 7.16 and their variants or a combination were used to translate *pachipachi*. More than half of the cases (41 cases, 63%) were translated using *crackle*, *pop*, or *sizzle* (or their variants).

It would be worth noting here that all expressions used to translate *pachipachi*, bar one (*explode*), are also onomatopoeia in English.

Gā

Next, *gā* is a phonomime that describes the sound of a hard object being shaven or a heavy object moving. It is also used to describe a vigorous manner of action. There are 48 uses of this expression in this corpus, and 45 random samples were examined.

At first glance, in sharp contrast with *pachipachi*, in most cases (37 cases, 77%) the translation of *gā* is omitted, and the remaining eight cases were translated as in Table 7.17.

Table 7.16 Frequently used target-text expressions for *pachipachi*

Frequency (*n*)	Target-text expression
15	crackle
17	pop
9	sizzle
4	fizz
4	spit
4	crack
3	spatter
2	crackle and pop
1	bubbling
1	crackle and sputter
1	explode
1	pop and sizzle
1	slapping
1	splatter
1	cracking and spit

Table 7.17 Frequently used target-text expressions for *gā*

Frequency (*n*)	Target-text expression
3	well
2	briefly
2	vigorously
1	paraphrase

Table 7.18 Frequently used target-text expressions for *puchipuchi*

Frequency (*n*)	Target-text expression
10	pop
3	grainy
3	lumpy
2	plump
1	crackling
1	crunchy
1	spongy

Puchipuchi

Puchipuchi is used to describe the sound of something small being popped, or of pressing a small button. In this corpus, it appears 39 times and 37 random samples were examined. One example had to be removed due to a concordance error. Only a limited range of expressions are used (Table 7.18) and, as with *gā*, at first glance *puchipuchi* seems to be mostly omitted from the translation.

Dōn

Dōn is an onomatopoeic expression (phonomime) that denotes the sound of objects or people crashing into something vigorously, or the dynamic manner of an action (Ono 2007). In this corpus it was used 23 times, and all occurrences were analysed in detail. One case was discarded due to a concordance error. At first glance, 11 cases out of 23 seemed to be omissions, while the rest were translated using expressions to denote a large quantity, such as *a lot*, *loads* or *tons* (Table 7.19).

Table 7.19 Frequently used target-text expressions for *dōn*

Frequency (*n*)	Target-text expression
2	a lot of it
2	all
1	boom
1	loads of
1	right in
1	thick and filling
1	tons
1	whole
1	dome-shaped

Table 7.20 Frequently used target-text expressions for *gangan*

Frequency (*n*)	Target-text expression
3	a lot of
3	a rolling boil
1	a ton of
1	all the way
1	intensely
1	vigorously
1	just
1	thoroughly

Gangan

Finally, *gangan* expresses a banging sound that occurs repeatedly, the way in which such a sound echoes internally in our bodies, or the vigorous and dynamic nature of an action (Ono 2007). There are 17 uses in this corpus, all of which were analysed in detail. Five cases were omissions, while the remaining 12 were translated using a number of expressions for vigorousness (Table 7.20).

So far, I have shown how each onomatopoeia is translated. Most onomatopoeia seem to be translated using a limited range of expressions. Table 7.21 provides a summary of the findings so far.

It is clear from Table 7.21 that there are highly established and consistent ways of translating onomatopoeia, especially those in Groups 1 and 2. Of all the onomatopoeia in

Table 7.21 The number of frequently used target-text expressions

Lema	Frequency (*n*)	Random sample (*n*)	Total no. of expressions used to translate (*n*)	No. of expressions used more than five times (*n*) (a)	Frequency of (a) (*n*)	Percentage of (a) in a random sample
ShiQkari	5018	536	75	10	343	63.99
TaQpuri	3124	504	59	8	380	75.40
ShiQtori	1669	442	24	4	390	88.24
SaQpari	939	366	29	5	320	87.43
Shinnari	927	365	14	5	342	93.70
Torotoro	3120	503	39	9	436	86.68
Fuwafuwa	3024	501	30	7	447	89.22
Sakusaku	1994	462	26	9	341	73.81
SaQ	1177	398	18	5	222	55.78
Karikari	1000	375	9	2	354	94.4
Pachipachi	73	65	12	3	41	63.08
Gā	48	45	4	0	0	0
Puchipuchi	39	37	7	1	10	27.02
Dōn	23	23	9	0	0	0
Gangan	17	17	8	0	0	0

Groups 1 and 2, eight expressions have more than 70% of their occurrences translated using a limited range of phrases. In fact, for six expressions over 85% of their occurrences are translated using a set number of phrases. These findings are somewhat different from Flyxe's argument that, as "the more highly lexicalized onomatopoeia only indirectly imitate sounds, they are more abstract and it is difficult to give a one-to-one translation" (Flyxe 2002, 70). Instead, the findings from this study so far indicate that for medium to highly established onomatopoeia, the translation is highly consistent. That is, in contrast to findings from previous studies, the current study has so far found somewhat consistent strategies to translate onomatopoeia, especially for Groups 1 and 2: most onomatopoeia in these categories were translated using one-to-one equivalence, regardless of the grammatical category.

The less lexicalised onomatopoeia in Group 3 present a somewhat different picture. Unlike expressions from Groups 1 and 2, onomatopoeia in this group do not seem to have a similar range of set phrases for translation. In fact, there are only four expressions that are used more than five times to translate the expressions in this group. Again, this is contrary to what Flyxe (2002) argues. Despite the fact that expressions from Group 3 seemingly have a stronger link to the sound, their translation is not as straightforward as the more abstract expressions.

It is worth noting here that the comparison between the percentage of frequent expressions and the percentage of omissions shows an interesting contrast (Table 7.22).

Earlier in the chapter I explained that some scholars reported omission cases: 5.9% reported by Flyxe (2002), 32.6% by Casas-Tost (2014), 19.3% by Inose (2008), or 16.3% reported by Minashima (2004). As shown in Table 7.22, omissions in Cookpad data seem to occur much less often than has been reported elsewhere, except for *saQ* from Group 2 and all of Group 3 onomatopoeia, apart from *pachipachi*, which has no cases of omissions. In the remainder of this chapter, I will focus on these five onomatopoeic expressions to identify translation challenges.

Table 7.22 The percentage of frequently used expressions and the percentage of omissions

Lema	Frequency (*n*)	A random sample (CI and CL based) (*n*)	Percentage of expressions used more than five times (*n*)	Omission cases (%)
ShiQkari	5018	536	63.9925373	18.20
TaQpuri	3124	504	75.3968254	7.97
ShiQtori	1669	442	88.2352941	4.99
SaQpari	939	366	87.431694	4.64
Shinnari	927	365	93.6986301	1.11
Torotoro	3120	503	86.6799205	2.61
Fuwafuwa	3024	501	89.2215569	0.48
Sakusaku	1994	462	73.8095238	18.40
SaQ	1177	398	55.7788945	36.36
Karikari	1000	375	94.4	2.94
Pachipachi	73	65	63.0769231	0
Gā	48	45	0	73.33
Puchipuchi	39	37	27.027027	47.22
Dōn	23	23	0	45.45
Gangan	17	17	0	29.41

3 Omission of Onomatopoeia

So far, I have shown how onomatopoeia is translated using one-to-one (or direct) equivalence. In contrast with findings from previous studies, the current study has found that strategies used for the translation of onomatopoeia seem to be somewhat consistent, especially for the more lexicalised onomatopoeia. However, focusing on cases of straightforward translation does not help us to identify the cause of the challenge, if any, faced by translators, and it is imperative to focus on cases where onomatopoeia is omitted from translation. After all, it was the high frequency of omission cases that led scholars to argue that the translation of onomatopoeia was a challenge for translators.

However, it is important to note that omission is an accepted translation strategy and that something being omitted from the target text should not necessarily mean that its translation is challenging. Indeed, Baker (1992, 40) argues that omission does not necessarily have a negative effect on translation:

Table 7.23 Summary of omissions

Lema	Consideration for Immediate Context	Consideration for Context	Consideration for Genre and Style
SaQ	114	28	
Gā	33	4	
Puchipuchi	4	13	
Dōn	3		6
Gangan	5		

> This strategy may sound rather drastic, but in fact it does no harm to omit translating a word or expression in some contexts. If the meaning conveyed by a particular item or expression is not vital enough to the development of the text to justify distracting the reader with lengthy explanations, translators can and often do simply omit translating the word or expression in question.

The view of omission as a legitimate translation strategy is also echoed by scholars such as Toury (1995) and Dimitriu (2004). Still, a close examination of cases of omission in the translation of onomatopoeia enabled me to shed light on the nature of onomatopoeia translation. This section therefore focuses on five expressions that demonstrate a relatively high number of omissions: *saQ*, *gā*, *puchipuchi*, *dōn*, and *gangan*. All the cases of omission from random samples were analysed, to identify why the onomatopoeia was not included in the translation. However, note that cases of omission within recipe titles are not included in the analysis, as it seems to be common practice to omit (parts of) referral recipe titles.

A close examination of omission cases indicates that there are roughly three categories of omissions (Table 7.23): (i) consideration for immediate context (or lexical consideration); (ii) consideration for context (existing assumptions); and (iii) consideration for genre and style. In the remainder of this chapter, I will examine a range of examples in each category.

3.1 Omission and Consideration for the Immediate (Lexical) Context

First, let us look at the case of *saQ*. At first glance this expression seems to be frequently omitted, with 147 cases of omissions (two cases were removed from the analysis as they were recipe titles), which amounts to

Table 7.24 Omission of onomatopoeia used with a range of verbs that indicate swiftness of action

	Original Japanese	Literal translation	Cookpad translation
1	もやし を さっと 茹で て 、水気 を 切り 、 one に 加え て 混ぜる 。	Parboil the bean sprouts *quickly*, strain and add Step 1 and mix evenly.	*Parboil* the bean sprouts, strain and add Step 1 and mix evenly.
2	仕上げ に ごま油大さじ one を 回しかけ て さっと 煽れ ば 完成!	Drizzle 1 tablespoon of sesame oil, toss *quickly*, and enjoy!	Drizzle 1 tablespoon of sesame oil, *toss*, and enjoy!
3	野菜 は それぞれ 食べやすい 大きさ に 切り 、 熱湯 で さっと ゆでる 。	Cut the vegetables up into easy to eat pieces, and blanch *quickly* in boiling water.	Cut the vegetables up into easy to eat pieces, and *blanch* in boiling water.

36% of the overall use. However, a close examination reveals that most of what appear to be omissions are in fact not. These are cases of cooccurrence with verbs that entail swiftness in English, such as *parboil*, *toss*, or *blanch* (Table 7.24).

In total, there are 114 cases such as these used with a range of verbs that indicate swiftness of action such as parboil or *toss*. This suggests that most omission cases are not in fact omissions, and that the impression that is communicated via the use of onomatopoeia is communicated as part of the verb phrase. That is, most omission cases of *saQ* seem to be the result of consideration for the immediate (lexical) context.

Omissions as a result of consideration for the immediate or lexical context are found across all five expressions that showed a relatively high frequency of omission. The omission of *gā* is particularly interesting, as an overwhelming 33 out of 45 cases seem to apply to this type of omission. However, a close examination of omissions reveals an interesting fact: namely, that most omission cases are, in fact, cases where the onomatopoeia *gā* is used as a metonymy for the use of a food processor or hand blender (Table 7.25).

An overwhelming 33 out of 37 cases (i.e. 89%) which appeared to be omissions are in fact not omissions. Instead, they turn out to be the cases in which the metonymic description in the original Japanese is paraphrased to spell out the use of a food processor. That is, in the original Japanese the onomatopoeia *gā*, which represents the sound of the food processor being

Table 7.25 Omission of onomatopoeia *gā* used as a metonymy for the use of a food processor

	Original Japanese	Literal translation	Cookpad translation
4	●をFPに入れてガーっとしておく。	Add ● in a food processor and make a grinding noise.	Add the ● marked ingredients into *a food processor and process* until fully combined.
5	ミキサーやフードプロセッサーがある場合は、全ての材料を入れて(アンチョビは1枚だけ)、ガーッとするだけ。	If you have a blender or food processor put all of the ingredients in it (just one anchovy) and make a grinding noise.	If you have a blender or food processor, put all of the ingredients into it (just one anchovy) and blend.

used, is now used as a metonymy to describe the whole process of using the food processor. Conversely, the use of the term *food processor* or *hand blender* in the target language typically enables readers to access its encyclopaedic information, which involves the noise. That is, rather than finding a formal equivalent for *gā* in English, the action of using the food processor is spelled out.

Other cases of omissions are where there is a standard English expression for the particular cooking process of piercing (or pricking) the surface of the meat, and where the Japanese *puchipuchi* is used in the original recipe (Table 7.26).

In these cases, while it may look as though the onomatopoeia was omitted, the fact is that it was not necessary as part of the cooking instruction, since the impression communicated via the use of *puchipuchi* in the original text is conveyed by the use of specific verbs.

Similar omission cases are found with the onomatopoeia *dōn* (Table 7.27).

The use of *dōn* typically gives rise to the impression of a dynamic and excited way of doing something—not a delicate and careful approach. The choice of verbs in these examples produces a similar effect.

Finally, *gangan* also has a range of omission cases that resulted from consideration for the immediate context (Table 7.28).

There are five omission cases, and in all of them the use of other constituents of the translation unit helps to give rise to effects similar to the impressions communicated via the use of onomatopoeia. That is, verbs such as *hit* or *beat* already indicate the vigorous nature of the action, and

Table 7.26 Omission of onomatopoeia *puchipuchi*

	Original Japanese	Literal translation	Cookpad translation
6	骨抜き完了。フォークで皮の方をプチプチと何箇所か刺しておきます。(多めに開けると、揚げた時に膨らんで爆発しそうになるのが防げます。)	The chicken wings are de-boned. *Pierce <u>in a prickling manner</u> the skin* side several times with a fork. (Piercing a lot of holes through the skin will prevent the chicken from bursting and splattering oil around while you're deep frying them.)	The chicken wings are de-boned. *<u>Pierce the skin</u>* side several times with a fork. (Piercing a lot of holes through the skin will prevent the chicken from bursting and splattering oil around while you're deep frying them.)
7	イカをさばき、(簡単な手順は【four】を参照)表面をプチプチ爪楊枝で刺して、数箇所穴を開ける	Clean the squid (refer to Step 4 for an easy way), and *pierce some holes <u>in a prickling manner</u>* in the surface using a toothpick.	Clean the squid (refer to Step 4 for an easy way), and *<u>pierce some holes</u>* in the surface using a toothpick.
8	以前は梅にぷちぷち穴あけて作ってましたが、この方法を友達に教わってからはこの方法で作ってます！風味がよくてすぐ飲んでしまう…。砂糖が溶け切った頃は風味と甘味、そのうち酸味が出てきて、最後に仁の風味が出てくるみたいです。お好みの所で止めてね！	I used to *prick holes <u>in a prickling manner</u>* on the skin of ume plums, but I've been making ume juice this way ever since my friend told me the technique! It's so delicious and fragrant that I drink it all up very quickly … You can enjoy the flavour and sweetness when the sugar has dissolved completely; it will be fragrant and sweet to start, then the syrup will develop the sourness. Lastly the slightly bitter "jin" flavour from the core of the ume will come out. Stop the ume juice from developing when it reaches the flavour that you like!	I used to *<u>prick holes</u>* on the skin of ume plums, but I've been making ume juice this way ever since my friend told me the technique! It's so delicious and fragrant that I drink it all up very quickly … You can enjoy the flavor and sweetness when the sugar has dissolved completely; it will be fragrant and sweet to start, then the syrup will develop the sourness. Lastly the slightly bitter "jin" flavor from the core of the ume will come out. Stop the ume juice from developing when it reaches the flavor that you like!

Table 7.27 Omission of onomatopoeia *dōn*

	Original Japanese	Literal translation	Cookpad translation
9	10分 ほど 置く と、 みょうが から の 水 分 と、 レモンポン 酢 が なじん で、 い い 感じ に なり ます 。 シンプル に、 冷 奴 に どーん と のせ る の も、 あり でし ょう ね。	If you let it sit for about 10 minutes, the moisture from the myoga ginger will combine with the lemon juice and ponzu quite nicely. It's simple and delicious *served dynamically over* cold tofu.	If you let it sit for about 10 minutes, the moisture from the myoga ginger will combine with the lemon juice and ponzu quite nicely. It's simple and delicious *served over* cold tofu.
10	入りきる 大きな お皿 が ない ので、 テー ブル の 上 に 土台 を 作っ て アルミホイル を 敷き、 ドーン と 置き まし た ★	I don't have a large serving dish, so I made a space on the table, put aluminium foil and *plopped it down dynamically*.	I don't have a large serving dish, so I made a space on the table, put aluminum foil and *plopped it down*.
11	ステーキ に どーん と かけ て、 どん と 食 べ て ね！人参 の グ ラッセ は、 レシピ ID: 2270274 を ご覧 ください ね！	*Pile these dynamically* next to your steak and devour them! Go to Recipe ID: 2270274 for tips on these glazed carrots.	*Pile these* next to your steak and devour them! Go to Recipe ID: 2270274 for tips on these glazed carrots.

as such the translation of the onomatopoeia is not essential for achieving relevance.

This analysis is in line with Inose's (2008) observation that the omission of onomatopoeia often happens when onomatopoeia are used as a modifier rather than carrying the essential information. She argues that onomatopoeia as modifiers "do not add new information to the phrase but elaborate, or even repeat the information already given by other words (verbs, for example)" (Inose 2008, 114). It is not entirely clear what she means by "do not add new information". However, as the examples given illustrate, the impression communicated via the use of onomatopoeia in the original Japanese is conveyed by other constituents of the translation unit. This makes a translation of onomatopoeia redundant.

So far, I have analysed cases of omission where lexical consideration resulted in the omission of onomatopoeic expressions from the translation. In the next section, I will turn to cases of omission where consideration for a wider context, or the context beyond the immediate (linguistic/lexical) context, results in what appears to be the omission of onomatopoeia.

Table 7.28 Omission of onomatopoeia *gangan*

	Original Japanese	Literal translation	Cookpad translation
13	肉たたきや包丁の背で牛肉の両面をガンガンと何度か叩く。	Use a meat hammer or the back of a knife to ***hit*** the meat several times ***vigorously***.	Use a meat hammer or the back of a knife ***to pound the meat*** several times.
14	もしミートハンマーがあったら、ガンガン叩いてくれ。	If you have a meat hammer, ***hit*** it ***vigorously***.	If you have a tenderizer, give the pork a few whacks.
15	とんがりコーンをビニール袋に入れて、棒でガンガン砕く。	Put the Tongari corn in a plastic bag and use a rolling pin ***to break them into pieces vigorously***.	Put the Tongari corn in a plastic bag and use a rolling pin to smash them.
16	細かくしたい時は、ガンガン砕いた後のゴリゴリ潰す作業を一生懸命やってください。フードプロセッサーや、すり鉢とすりこ木を使うのもいいかも。すぐにしけってしまいそうなので、必要な分をその都度、作った方がいいと思います。	When crushing up into pieces, hit first ***viciously***, then roll to crush up. You can use a food processor if you like, or mortar and pestle. It's best to use it right away, so make what you need.	When crushing up into pieces, hit first, then roll to crush up. You can use a food processor if you like, or mortar and pestle. It's best to use it right away, so make what you need.
17	■注意点■メレンゲの見極めですが時間で表記は目安です、卵白だけのメレンゲのような固さにまでガンガンにしない事、焼いたらボソッとなる原因です湯煎無しでも泡立てばそれでも可です●焼き環境も目安です、自宅のオーブンの取説と比べて判断してください。	■ Attention: ■ Regarding the beating times indicated for the egg, please note that they are just guidelines. Make sure that you do not beat it ***vigorously*** as stiff as you would an egg white meringue, or the sponge will be very dry. You can also try beating the egg at room temperature rather than holding it over a hot water bath. The baking time is also approximate; please adjust it depending on your oven, observing how well or fast the cake is baking.	■ Attention: ■ Regarding the beating times indicated for the egg, please note that they are just a guideline. Make sure that you do not beat it as stiff as you would an egg white meringue, or the sponge will be very dry. You can also try beating the egg at room temperature rather than held over a hot water bath. The baking time is also approximate; please adjust it depending on your oven, observing how well or fast the cake is baking.

3.2 Omission and Consideration for Context

In the previous section, I showed how other (lexical) constituents of the translation unit may give rise to an impression which is communicated via the use of onomatopoeia in the original Japanese. In this section, I will show how consideration of a context beyond the immediate constituents of the translation unit could result in omission.

I previously argued that the majority of omission cases for *saQ* can be explained in terms of consideration for the immediate context. The remainder of omission cases for *saQ* require consideration for the context beyond the lexical context (Table 7.29).

Table 7.29 Omission of onomatopoeia *saQ* as a result of contextual consideration

	Original Japanese	Literal translation	Cookpad translation
18	蓮根 は スライスし たら 、変色 を 防ぐ ため に サッと 水 に さらし て ください 。 できる だけ 薄く スライスす る の が おいしさ の ポイント です 。 蓮根 は 半透明 に なっ たら ザル に あげ ます 。 ゆで過ぎる と シャキ シャキ 感 が なく なっ て しまう ので 注意し ましょう 。	Soak the lotus root *swiftly* in water after slicing it to prevent it from changing colour. It's best to slice the lotus root as thinly as possible. Drain the lotus root as soon as it turns semi-translucent. If you boil it too much it'll lose its crispiness so be careful.	Soak the lotus root in water after slicing it to prevent it from changing colour. It's best to slice the lotus root as thinly as possible. Drain the lotus root as soon as it turns semi-translucent. If you boil it too much it'll lose its crispiness so be careful.
19	チンゲン菜 は 葉 を ほ ぐし て シンナリする 程度 に さっと 茹でる 。 冷水 に 取っ て 水 気 を 切っ て おく 。	Separate the bok choy leaves, and boil *quickly* until they wilt. Chill with cold water, and drain.	Separate the bok choy leaves, and boil until they wilt. Chill with cold water, and drain.
20	お客様 が 来る 前 に 、 おうち に ある 物 で さっと 1品 作り たく て 思いつい た レシピ ♪	I created this recipe *to cook quickly* with leftovers before a friend came to visit me at home.	I created this recipe with leftovers before a friend came to visit me at home.
21	豆苗 は 半分 に 切っ て 熱湯 に さっと 潜らせ (30秒程度) 、 ザル に あげる 。	Cut the pea shoots in half, *quickly* soak in hot water for 30 seconds, then drain.	Cut the pea shoots in half, soak in hot water for 30 seconds, then drain.

In these examples, the omission of onomatopoeia can be explained as a result of contextual consideration: for example, in (18) in the context in which the need to soak the ingredients after slicing to prevent discolouring would guide the hearer to recover assumptions such as that the ingredients need soaking immediately; and similarly, in (19) in the fact that the outcome of the particular process (to boil in this case) would lead the hearer to access assumptions as to how quickly leafy vegetables wilt when added to boiling water. Example (20) is not directly related to cooking processes. However, the context given for this recipe, which involves hosting a friend at home, would lead to the recovery of an assumption that the author would want to prepare the recipe quickly. Finally, in (21) the cooking duration is already specified as 30 seconds, making it unnecessary to specify how quickly the pea shoots need to be cooked using onomatopoeia.

This type of omission can be seen across expressions. For example, four remaining cases of omission of *gā*, which describes a vigorous manner of action, can be explained either because the vigorous manner can be inferred from the context, or the time of cooking is being specified. As a result, the onomatopoeia is in fact redundant when conveying essential information only (Table 7.30).

Next, at 47.22%, almost half of the uses of *puchipuchi* appeared to be omitted from the translation. Out of 17 omission cases, 13 cases seem to be the result of consideration for contextual assumptions, especially when referring to the texture of an object, and are omitted as readers are expected to know what it feels like to eat certain foods (Table 7.31).

So far, I have shown that omission cases can be accounted for in terms of consideration for the (immediate) context, which allows for the communication of impressions in the target text, and which would have been communicated via the use of onomatopoeia in the original Japanese. Whether it resulted from consideration for the immediate or lexical context of the translation unit, or from consideration for the context beyond other constituents of the translation unit, the impression communicated via the use of onomatopoeia in the original Japanese is, in fact, conveyed by other assumptions (made) available in the context in which the onomatopoeia is used (or omitted). In the next section, I will examine a range of omission cases that do not fit into either category.

Table 7.30 Omission of onomatopoeia *gā* as a result of contextual consideration

	Original Japanese	Literal translation	Cookpad translation
22	もちろん ボール で ガーッと 混ぜ て も 作れ ます 。 12個取り の 焼きドーナツ型 、 小さい カップケーキ型 は フリーザーパック使用 が お勧め です 。甘さ は 控えめ です 。 好み で 増やし て ください 。	Of course, you can still make this by mixing *vigorously* everything together in a bowl instead. I recommend using a freezer bag if you're just making enough batter for a 12 donut mold or small cupcakes though. This recipe isn't so sweet so if you want to add more sugar, feel free.	Of course, you can still make this by mixing everything together in a bowl instead. I recommend using a freezer bag if you're just making enough batter for a 12 donut mold or small cupcakes though. This recipe isn't so sweet so if you want to add more sugar, feel free.
23	ボウル に 卵・薄力粉 を 入れ て よく 混ぜる 。 白っぽく ムラ なく 混ざ る と O K 。 泡立て器 で 手早く 一気 に ガーッと やる の が コツ 。	Combine the eggs and flour in a bowl and mix well. When it becomes white and free of lumps, it's ready. Quickly mix it *vigorously* together with a whisk.	Combine the eggs and flour in a bowl and mix well. When it becomes white and free of lumps, it's ready. Quickly mix it together with a whisk.
24	フライパン を 強火 で ガーっ と 熱す るこれ が ポイント です!	Heat a pan *vigorously* over high heat. This part is important!	Heat a pan over high heat. This part is important!
25	じゃがいも を 調味料ごと 入れ 、 酒 、 みりん を 加え 、 フタ を し て ガーッと 10分 加熱し ます 。	Add the potatoes along with the seasoning. Add the sake and mirin. Cover and cook *vigorously* for 10 minutes.	Add the potatoes along with the seasoning. Add the sake and mirin. Cover and cook for 10 minutes.

Table 7.31 Omission of onomatopoeia *puchipuchi* as a result of contextual consideration

	Original Japanese	Literal translation	Cookpad translation
26	ぷちぷち コーン 入り の もっちもち蒸しパン 。 軽食 に ぴったり の 塩味タイプ 。 ヘルシー な ノンオイル・ノンエッグ 。	This chewy bread gets its <u>*popping*</u> texture from the corn kernels that are added in. This is a perfectly light savoury snack. It has oatmeal, which is healthy for you, and there are no eggs.	This chewy bread gets its great texture from the corn kernels that are added in. This is a perfectly light savory snack. It has oatmeal, which is healthy for you, and there are no eggs.
27	ぷちぷち 食感 が 楽しい キヌア を 美味しく サラダ で 食べ たく て 。	I wanted to eat a delicious salad to enjoy the fun <u>*popping*</u> texture of quinoa.	I wanted to eat a delicious salad to enjoy the fun texture of quinoa.
28	私 は 甘め が 好き な ので 、 砂糖多め です ! お好み で 調節し て 下さい ね ♡ゴマ の プチプチし た 食感 も 残し たい ので 、 半擂り に し ます 。 日持ち日数 は どの くらい でしょう か 、 私 は いつ も 保存ビン (写真) に いれ て 冷蔵庫 で 保存 、 1 週間 ~ 10日 で 使い切り ます 。	I like this to be on the sweet side, so I've added quite a lot of sugar! Please adjust to your taste. I wanted the <u>*popping*</u> texture of the sesame seeds to remain, so I just ground them up halfway. I am not sure how long this will keep. I always store it in a jar (see photo) in the refrigerator, and use it up within a week to 10 days.	I like this to be on the sweet side, so I've added quite a lot of sugar! Please adjust to your taste. I wanted the texture of the sesame seeds to remain, so I just ground them up halfway. I am not sure how long this will keep. I always store it in a jar (see photo) in the refrigerator, and use it up within a week to 10 days.
29	手作りいくら と 鮭 の 親子丼 です。 鮭 の プチプチ 感 が 堪ら ない とても 美味しい 丼 です 。	Here's a recipe to make oyakodon (parent-and-child rice bowl) with homemade ikura salmon roe and salmon. The pretty colour and characteristic <u>*popping*</u> texture of the ikura is irresistible.	Here's a recipe to make oyakodon (parent-and-child rice bowl) with homemade ikura salmon roe and salmon. The pretty color and characteristic texture of the ikura is irresistible.

3.3 Omission and Consideration for Genre and Style

The previous sections illustrated how instances of omission can be explained in terms of consideration for the immediate (or lexical) context or consideration for context beyond lexical constituents. However, there are some cases of omission that do not fall into either category (Table 7.32).

These cases all involve the omission of *dōn*. As these examples show, the underlined expressions in the literal translation add little information in English and seem somewhat out of place. In most cases, the onomatopoeia in the literal translation appears to add dramatic effect to the action, providing a further description of what has already been described, rather than being part of the cooking process, as if to make a personal appeal to readers or to speak to readers face-to-face. This is probably linked to the strong social media aspect of Cookpad, as described earlier. Cookpad is a recipe site uploaded by users, and each user has a profile. Some established users have followers who comment on the recipe, to which the user can also reply. This social media-like function of Cookpad makes the use of *dōn*, as described earlier, acceptable, as it allows for and expects a certain amount of interaction between users. However, the social media aspect of Cookpad is yet to develop in non-Japanese sites and, in order for the target-language reader to understand, they would need to search for the relevance of such an addition. The omission of such onomatopoeia might result in the loss of communication of the impression, but the translated text would be easier for the target-text audience to follow. Indeed, the omission of onomatopoeia does not necessarily lead to a lack of information that hinders readers from following the recipe. That is, while the omission of onomatopoeia might result in the loss of an impression that was communicated via its use in the source language, the text still conveys the necessary information to function as a recipe.

Of course, this does not mean that the dramatic effects communicated in the original via the use of onomatopoeia are not important. Otherwise, why would the original author include them in the first place? However, the omission of onomatopoeia from the translation would allow for the avoidance of extra processing effort to recover what is essentially unnecessary. This would suggest that the decision to omit onomatopoeia in these cases might be motivated by careful consideration in order not to impose

Table 7.32 Omission of onomatopoeia as a result of considerations of genre and style

	Original Japanese	Literal translation	Cookpad translation
30	絶品★ 骨付きチキン が ドーン と、お口 で ほろほろ に とろけ ます♪旨み が 凝縮さ れ た 簡単&本格的チキンカレー♥	*Boom!* this delicious bone-in chicken melts in your mouth. It's an easy, yet authentic and flavourful chicken curry.	This delicious, gorgeous bone-in chicken thigh will melt in your mouth. It's an easy, yet authentic and flavorful chicken curry.
31	ドーン！しっかり 焼き色 が つい て いる の に、\n中 は フワフワ～♪翌日 も フワフワ～♪\n 甘く て 美味しい パン でき ました ぁ～!!	*Boom!* With a nice golden brown crust and a fluffy inside. This bread is fluffy even the following day. Enjoy this sweet and delicious bread!!	With a nice golden brown crust and a fluffy inside. This bread is fluffy even the following day. Enjoy this sweet and delicious bread!!
32	この ビーフ 1枚 を 広げ、大葉、ドレッシング の 玉葱、梅干少々 を 乗せ て くるり と 巻い て 半分 に 切る。皿 に 盛り付ける。ドレッシング液体 を たっぷり 切り口 から かけ て これ を 冷蔵庫 で ドーン と 冷やし 召し上がれ♪	Spread out one slice of the meat and lay a shiso leaf, the spring onion dressing, and a bit of pickled plum on top, roll it up, and cut it in half. Arrange onto plates. Drizzle lots of dressing over the openings of the meat, let cool in the fridge, *boom!* and enjoy.	Spread out one slice of the meat and lay a shiso leaf, the spring onion dressing, and a bit of pickled plum on top, roll it up, and cut it in half. Arrange onto plates. Drizzle lots of dressing over the openings of the meat, let cool in the fridge, and enjoy.
33	できあがり。玉ねぎ は 溶け て なくなっ ちゃっ てい ます。大きな チキン を ドーン と メイン に 盛り付け て ください。	It's done. The onion should be completely cooked down. Arrange the chicken thigh, *boom!*, in the centre of the plate when serving.	It's done. The onion should be completely cooked down. Arrange the chicken thigh in the center of the plate when serving.
34	生米 を 使う 場合 は、ココ で 入れ 炒める (研が なく て OK)(゜∀゜)o 彡 その まま ドーン＊ご飯 を 使う 場合 は まだ 入れ ません	If using uncooked rice add it at this point and cook (you don't need to rinse). *Put it right in!* If using cooked rice do not add it at this point.	If using uncooked rice add it at this point and cook (You don't need to rinse). If using cooked rice do not add it at this point.

unnecessary processing effort. That is, the loss of extra effects communicated in the original is balanced out by requiring less processing effort for the target-text readers.

4 Discussion and Conclusion

This chapter began by reviewing how the translation of onomatopoeia has been discussed in previous studies. It became apparent that there is an underlying assumption among scholars in Translation Studies that onomatopoeia is a challenge for translators. This assumption seems to motivate scholars to take a taxonomic approach, where the focus of their studies is to classify strategies and techniques used to translate onomatopoeia. Many scholars also note that onomatopoeia is often left untranslated or omitted.

Scholars have also discussed why onomatopoeia is a challenge for translators. Some argue that the difference is due to the phonetic systems between languages (Bartashova and Sichinskiy 2014; Flyxe 2002), and others blame the lack of one-to-one equivalence (Inose 2008), differences in the register or style of onomatopoeia use (Flyxe 2002), or unfamiliarity with onomatopoeia on the part of translators (Minashima 2004).

However, there is no evidence-based research to show that onomatopoeia is indeed a challenge for translators. It is true that there is often a lack of one-to-one equivalence when translating onomatopoeia. As a number of previous studies have shown, expressions across different parts of speech and grammatical classes are used to translate onomatopoeia. This has led scholars to conclude that onomatopoeia is a challenge for translation, when in fact there is little empirical evidence that it is indeed so. The motivation for this chapter, therefore, was to address this gap in the literature and to examine whether onomatopoeia is really difficult to translate.

Using bilingual data from Cookpad, a parallel corpus was built. Cookpad contained over 200 onomatopoeic expressions and their variants. All expressions were categorised into three groups, using Flyxe's classification developed from Tamori and Schorup's (1999) analysis. Group 1 included highly lexicalised onomatopoeia, which could be used without the quotative particle *to*, and consisted largely of psychomimes. Group 2 included onomatopoeia for which the use of the quotative *to* was optional and which were mainly phenomimes. Finally, Group 3 consisted of the least lexicalised onomatopoeia, mostly phonomimes, where the quotative *to* was compulsory. For this study, the five most frequent expressions from each category were analysed, together with random samples of their concordance.

Even given these limits, this study sheds some light on a myth about onomatopoeia translation. Namely, contrary to the consensus in previous studies, onomatopoeia does not seem to impose many challenges for translation. It is true that there are many expressions used to translate each onomatopoeia. However, a close examination reveals that only a limited number of expressions are indeed used repeatedly. In fact, in more than half (8 out of 15) expressions, over 70% of all instances are translated using a limited number of translations, with the highest percentage being 94.4% and the lowest being 73.8%. That is, for these eight expressions, an average of 86.109% instances were translated using a limited number of set phrases. It is interesting that these eight expressions are from Groups 1 and 2. This suggests that the translation of more lexicalised onomatopoeia is stable, contrary to what Flyxe (2002) claims. Furthermore, the fact that there was a large number of possible expressions to use in translation, but only a limited number out of all possible candidates were used repeatedly, indicates that the translation of onomatopoeia is much more stabilised and straightforward (as Inose puts it) than has previously been thought.

In order to identify why other onomatopoeia, especially those from Group 3, behaved differently, the instances of omission were examined carefully for the five expressions that had a high omission rate. A careful examination of each omission instance revealed that onomatopoeia in these cases was not simply omitted from translation because it was difficult to translate. It became apparent that what appeared to be instances of omission were in fact the result of considerations of relevance. In some cases, it was not necessary to include the onomatopoeia in the translation, as impressions were communicated via the other constituents of the translation unit, or they were inferable from the context beyond the immediate situation.

In other cases, consideration for writing style seemed to have influenced what appears to be an omission. In all cases, the inclusion of onomatopoeia was not necessary. In fact, inclusion might have imposed extra processing effort without worthwhile reward. That is, the decision to omit onomatopoeia was motivated by careful consideration for avoiding unnecessary processing effort. The loss of extra effects communicated in the original is balanced out by requiring less processing effort on the part of target-text readers.

This suggests that Inose (2008) was partially accurate in the suggestion that the omission of onomatopoeia often occurs when it functions as a modifier, as it does not add new (or essential) information. However, this study shows another dimension of omission—as most cases of omission examined in this chapter demonstrate, the impressions communicated via the use of onomatopoeia in the original Japanese are delivered by other means, and as such the translation of onomatopoeia is not entirely necessary.

Of course, the nature of onomatopoeia means that its translation is highly context bound and thus there seem to be a limited number of established techniques. As has been discussed throughout this book, onomatopoeia is deeply rooted in human sensory experience. Consequently, it can be extremely difficult to put into words. In addition, as frequently discussed in sound-symbolism approaches, different languages have different morphological and phonological constraints. However, as the careful examination of each expression in this chapter has shown, the lack of direct equivalence does not necessarily result in difficulties in translation. The combination of these factors is what has led to the perceived difficulties. Nevertheless, the analysis presented in this chapter has shown that it is consideration of the context in which the nebulous effects of onomatopoeia use are communicated that explains why and how the effect of onomatopoeia is translated, rather than a failed attempt to find linguistic equivalence.

Bibliography

Baker, Mona. 1992. *In Other Words: A Coursebook on Translation*. London: Routledge.

Bartashova, Olga A., and Anton E. Sichinskiy. 2014. Japanese–English Onomatopoeic and Mimetic Parallels: The Problem of Translatability. *Journal of Siberian Federal University. Humanities & Social Sciences* 2: 222–229.

Casas-Tost, Helena. 2014. Translating Onomatopoeia from Chinese into Spanish: A Corpus-Based Analysis. *Perspectives* 22 (1): 39–55. https://doi.org/10.1080/0907676X.2012.712144.

Dimitriu, Rodica. 2004. Omission in Translation. *Perspectives* 12 (3): 163–175. https://doi.org/10.1080/0907676X.2004.9961499.

Edström, Bert. 1989. Japanese Onomatopoetic Words: A Research Note. *Orientaliska Studier* 65: 35–52.

Flyxe, Martin. 2002. Translation of Japanese Onomatopoeia Into Swedish (with Focus on Lexicalization). *Africa & Asia* 2: 54–73.

Harashima, Jun, Michiaki Ariga, Kenta Murata, and Masayuki Ioki. 2016. A Large-Scale Recipe and Meal Data Collection as Infrastructure for Food Research. In *Proceedings of the 10th International Conference on Language Resources and Evaluation*, 2455–2459. https://www.semanticscholar.org/paper/A-Large-scale-Recipe-and-Meal-Data-Collection-as-Harashima-Ariga/8660210d71d56c480bd568f49fcf5243ec80a84f.

Hayase, Mitsuaki. 1978. Characteristics of Japanese Onomatopoeia Observed Through the English Translation. *Kanssai Gaikokugodaigaku Kenkyu Ronshu* 28: 117–127.

Inose, Hiroko. 2008. Translating Japanese onomatopoeia and mimetic words. In *Translation Research Projects*, ed. Anthony Pym and Alexander Perekrestenko, vol. 1, 97–116. Tarragona: Intercultural Studies Group.

Kubo, Atsuko. 1995. Miyazawa Kenji No Onomatope No Sekai [The World of Onomatopoeia in Miyazawa Kenji's Work]. *Kobe Kaisei Review* 130 (34): 17–30.

Marshall, Shelly. 2014. *Night on the Milky Way Railroad* (Kindle Edition). Los Gatos: Smashwords.

Minashima, Hiroshi. 2004. Nihhongo No Onomatope [Onomatopoeia in Japanese]. *Fukui Daigaku Kyoiku Chiiki Kagakubu Kiyou* 60: 97–115.

Molina, Lucia. 2006. *El Otoño Del Pingüino. Análisis Descriptivo de La Traducción de Los Culturemas*. Castelló de la Plana: Publicacions de la Universitat Jaume I.

Murakami, Haruki. 1999. *Suputoniku no Koibito [Sputnik Sweeaheart]*. Tokyo: Kodansha.

Ono, Masahiro. 2007. *Nihongo Onomatope Jiten [Japanese Onomatopoeia Dictionary]*. Tokyo: Shogakukan Publishing.

Sato, Eriko. 2017. Translanguaging in Translation: Evidence from Japanese Mimetics. *International Journal of Linguistics and Communication*. 5. https://doi.org/10.15640/ijlc.v5n1a2.

Takeuchi, K. 1998. Schôchôgo no honyaku o meguru futatsuno mondai [Two Issues on the Translation of Symbolic Expressions]. *Musashi daigaku jinbun-gakkai zasshi* 24 (4): 47–77.

Tamori, Ikuhiro, and Lawrence Clifford Schourup. 1999. *Onomatope: Keitai-to Imi [Onomatopoeia: Form and Meaning]*. Tokyo: Kuroshio Shuppan.

Toury, Gideon. 1995. *Descriptive Translation Studies – And Beyond*. Amsterdam and Philadelphia: John Benjamins.

———. 2004. *Los Estudios Descriptivos de Traducción y Más Allá. Metodología de La Investigación En Estudios de Traducción*. Madrid: Catedra. https://www.catedra.com/libro.php?codigo_comercial=130064.

8

Conclusion

The aim of this study was to develop existing work in relevance theory on non-verbal communication and expressive meaning to account for onomatopoeia, which is somewhat under-researched from the perspective of communication, in contrast to the rich bodies of research concerning the sound-symbolic aspects of onomatopoeia. Onomatopoeia is often considered to be located at the edge of language and hence deeply connected to sensory experience. As a result, onomatopoeia research has been dominated by studies that are concerned largely with onomatopoeia's iconicity or sound symbolism; that is, the link between sound and meaning. However, as I showed in Chap. 2, context plays a significant role in interpreting onomatopoeia in communication. For example, while a single onomatopoeia might appear to communicate a variety of meanings, more than one expression could communicate something very similar. In addition, I showed that the contextual assumptions to which the hearer has access could also lead to a difference in interpretation. This raises the question of what and how onomatopoeia communicates, and how the use of onomatopoeia interacts with the context in which it is used.

Furthermore, as is generally acknowledged, the degree of mimeticity, or the degree of lexicalisation, of onomatopoeia varies. Some onomatopoeia

R. Sasamoto, *Onomatopoeia and Relevance*, Palgrave Studies in Sound,
https://doi.org/10.1007/978-3-030-26318-8_8

are highly conventionalised and therefore we may no longer notice that they are onomatopoeia (e.g. *chat*), while others are highly novel and could even be used once (e.g. *gwash*). This raises the question of what onomatopoeia encodes—if there is any coded element—and how onomatopoeia as a lexical item (or not) is used in communication. Against this backdrop, this study was an attempt to examine the role of onomatopoeia in communication. In particular, drawing on existing work within relevance theory, the study aimed to propose an explanatory account of the role of onomatopoeia in communication as part of ordinary language use, without calling for special treatment.

To that end, I have drawn upon the relevance-theoretic notions of the *showing–saying* continuum (Sperber and Wilson 1986/1995; Wilson and Wharton 2006; Wharton 2009) and impression (Sperber and Wilson (2015) and have argued that onomatopoeia necessarily involves both *showing* and *saying* aspects of communication, providing both direct (non-coded) and indirect (coded) evidence for ostension. As I discussed in detail in Chap. 3, the direct evidence that the speaker provides in using onomatopoeia is based on perceptual resemblance. That is, onomatopoeia enables the speaker to provide a verbal (coded) cue via its encoded content, while allowing them to present what perceptually resembles the experience via its phonological form. For example, the use of the onomatopoeia *ting* provides indirect evidence for communication delivered via the encoded concept of *the sound emitted by a small bell* (*OED*), while *showing* provides direct, non-coded evidence for what the sound of the bell is like via its phonological form which resembles the experience. I then turned, in Chap. 4, to the coded aspect of onomatopoeia and explained how what appears to be the polysemous nature of onomatopoeia can be accounted for in terms of ad hoc concepts—a notion developed in relevance-theoretic lexical pragmatics. I argued that rather than being polysemous, the encoded concept of an onomatopoeia is adjusted (broadened or narrowed, or both) to suit a specific context. What appears to be the polysemous nature of onomatopoeia is therefore a result of pragmatic processing, rather than semantic encoding, in context.

The implication of this study is that what appears to be a systematic link between sound and meaning is a result of human inferential processes. As mentioned earlier, the huge body of research on onomatopoeia has shown that humans are capable of making links between sound and meaning. However, my point is that such links do not exist independently of the human mind. Rather, humans create such links. That is, onomatopoeia is humans' attempt to "recreate" their perception of their sensory experience, a language-based tool to share impressions with each other.

As mentioned in Chap. 1, there has been some debate about terminology. Scholars have used terms such as *onomatopoeia*, *mimetics*, *phonomime*, *phenomime*, *psychomime*, *ideophone*, and so on. The issue with terminology particularly pertains to the debate about what onomatopoeia actually is. However, as is generally acknowledged, onomatopoeia ranges from novel and creative examples to highly lexicalised ones. Furthermore, some onomatopoeia can be used to describe sensory experience that does not pertain to one particular modality but involves cross-modal senses. By positioning onomatopoeia as a device that humans use to *show* and *say* simultaneously, rather than relying on what appears to be a systematic link between sound and meaning, this study illustrates how the terminological divide is not necessary, in that onomatopoeia is a *property* of a linguistic form, rather than a word class. This is further supported by the existence of the quasi-onomatopoeia discussed in Chap. 4, which is an onomatopoeia-like use of forms that do not originate from onomatopoeia. The implication of this is that whether one chooses *onomatopoeia* or *mimetics*, or whether these expressions are sound or quality based, we can account for the role they play in communication in the same way, thus making the terminological conflict somewhat unnecessary, at least in the discussion of their role in communication.

This study is not the first to claim that onomatopoeia communicates impressions. In many studies it is often implicitly assumed that onomatopoeia involves the communication of impressions and allows for "re-experience". However, most often, terms such as *impressions* or *re-experience* are used in the ordinary sense of the word, without providing further explanation. In contrast, as a cognitively grounded theory, relevance theory provides an explanatory architecture for all aspects of human communication and, as such, allows for the discussion of how such *impressions* can be communicated and what it means to "share impressions" or "re-experience", beyond the ordinary uses of these terms.

This analysis also has implications for relevance theory. As I discussed in Chap. 3, most cases of *showing* in most of the relevance-theoretic literature are based on visual evidence, such as showing a bandaged wrist to communicate that the speaker cannot play tennis (Sperber and Wilson 2015), facial or bodily expressions such as a smile or a gaze (Wharton 2009), or metalinguistic resemblance (e.g. Sperber and Wilson 1986/1995; Wilson and Sperber 2012). This study analysed onomatopoeia in terms of the *showing–saying* continuum. The implication is that onomatopoeia, which has a linguistic element, also has a *showing* aspect, which has been somewhat under-researched. This study therefore paves the way for a possible new direction in the discussion of the *showing–saying* continuum and verbal versus non-verbal communication. Furthermore, I have argued that the use of onomatopoeia involves providing direct (non-coded) evidence by virtue of *perceptual resemblance*. To my knowledge, the discussion on resemblance in relevance theory has been concerned mainly with *interpretive resemblance*, rather than *perceptual resemblance*. This is probably because onomatopoeia has attracted little attention in pragmatics in general. This study, therefore, could open up an interesting discussion on the nature of resemblance.

In the second half of this book, I explored different aspects of onomatopoeia use: onomatopoeia in food discourse, onomatopoeia and Japanese culture, and onomatopoeia and translation. In each chapter, I presented how the relevance-theoretic account explains the role of onomatopoeia in this discourse. As discussed in Chap. 5, food writing is not just about describing food. Instead, it needs to evoke impressions and make readers relive the experience of food. Furthermore, food experience is not only about taste: all aspects of eating, including taste, texture, smell, and even the setting itself, matter in food experience. That is, food experience is itself multi- and cross-modal. For this reason, it is not surprising that onomatopoeia is prevalent in food discourse. The nature of onomatopoeia, which necessarily involves the sharing of impressions beyond one modality, provides an ideal tool for writers to convey their sensory experiences involving food.

The visual nature of Japanese culture, as well as the prevalence of onomatopoeia in the Japanese language, are widely recognised. In Chap. 6 I examined a range of onomatopoeia uses in manga, which is the prime example of the multimodal and visual nature of Japanese culture. Drawing on previous work on argumentation and manga (Rohan et al. 2018), I argued how onomatopoeia brings together multiple inputs in manga, creating a further impression of the scene and reinforcing manga's persuasiveness. This was an attempt to account for an aspect of a culture in terms of pragmatics. While sound symbolism is considered universal, onomatopoeia is often culturally bound. This analysis could therefore lead to further discussion of how such culturally bound items can be explained in universal terms from a communicative perspective.

Finally, in Chap. 7 I focused on the translation of onomatopoeia using user-generated data from Cookpad. While the corpus-based approach had already been taken in the analysis of the translation of onomatopoeia, the focus had been on literary texts written by professional authors and translated by professional translators, with little work being done using user-generated data. This study, therefore, is an attempt to examine ordinary language use, employing user-generated data. In Translation Studies, it is generally agreed that translating onomatopoeia is a challenge for translators. Previous studies are therefore focused mainly on describing translation strategies for onomatopoeia, and often claim that onomatopoeia is largely omitted. In contrast to these descriptive studies, I examined samples from the Cookpad corpus and argued that translation strategies for onomatopoeia are in fact consistent, and that omission is not a result of the difficulties of translating onomatopoeia. Instead, I examined a range of omission cases from the perspective of relevance theory, and argued that the omission of onomatopoeia from the target text is a result of considerations of relevance, and that its inclusion would have caused unjustifiable processing effort.

Throughout this study, the focus was solely on the role of onomatopoeia in communication, and the analysis went beyond the classification and description of linguistic features. While this allowed for the discussion of onomatopoeia in general, little attention has been paid to the linguistic and structural aspects of onomatopoeia. Furthermore, there was scant discussion of how the proposed account of onomatopoeia would link with other stylistic and expressive aspects of pragmatics, including pragmatic repetition (and duplication), as well as metaphor and metonymy. Furthermore, this study examined onomatopoeia only in

English and Japanese. The examination of examples from a broader range of languages would provide further evidence for the role of onomatopoeia in communication. Finally, while the only genre discussed in detail in this study was food discourse, there is a whole range of genres that would involve the communication of impressions and sensory experience. Such genres would include travel writing, music discourse, dance, and even sports instructions. These await further research.

Bibliography

Rohan, Olivia, Ryoko Sasamoto, and Rebecca Jackson. 2018. Argumentation, Relevance Theory and Persuasion: An Analysis of Onomatopoeia in Japanese Publications Using Manga Stylistics. *International Review of Pragmatics* 10 (2): 219–242.

Sperber, Dan, and Deirdre Wilson. 1986/1995. *Relevance: Communication and Cognition*, 2nd ed. Oxford: Blackwell.

———. 2015. Beyond Speaker's Meaning. *Croatian Journal of Philosophy* 15 (2): 117–149.

Wharton, Tim. 2009. *Pragmatics and Non-Verbal Communication*. Cambridge: Cambridge University Press. https://doi.org/10.1017/CBO9780511635649.

Wilson, Deirdre, and Dan Sperber. 2012. *Meaning and Relevance*. Cambridge: Cambridge University Press. https://www.cambridge.org/core/books/meaning-and-relevance/500D9D857CB3274BD23AF6FB381133B7.

Wilson, Deirdre, and Tim Wharton. 2006. Relevance and Prosody. *Journal of Pragmatics, Special Issue: Prosody and Pragmatics* 38 (10): 1559–1579. https://doi.org/10.1016/j.pragma.2005.04.012.

Bibliography

Abdul Razak, Siti Hajjar. 2012. Japanese Onomatopoeia: Cosmetics and Skincare Products in Fashion Magazines. *Center for Comparative Japanese Studies Annual Bulletin* 8: 123–128.

Ahlner, Felix, and Jordan Zlatev. 2010. Cross-Modal Iconicity: A Cognitive Semiotic Approach to Sound Symbolism. *Sign Systems Studies* 38 (1/4): 298–348.

Akita, Kimi. 2009. *A Grammar of Sound-Symbolic Words in Japanese: Theoretical Approaches to Iconic and Lexical Properties of Mimetics*. Kobe: Kobe University.

———. 2013a. Kyoki Tokusei Kara Miru Onomatope No Fureemu Riron. In *Onomatope Kenkyu No Shatei: Chikazuku Oto to Imi*, ed. Kazuko Shinohara and Ryoko Uno, 101–116. Tokyo: Hitsuji Shobo.

———. 2013b. Onomatope/Otoshocho No Kenkyu-Shi. In *Onomatope Kenkyu No Shatei: Chikazuku Oto to Imi*, ed. Kazuko Shinohara and Ryoko Uno, 333–364. Tokyo: Hitsuji Shobo.

———. 2013c. Constraints on the Semantic Extension of Onomatopoeia. *Public Journal of Semiotics* 5 (1): 21–37.

———. 2017. The Linguistic Integration of Japanese Ideophones and Its Typological Implications. *Canadian Journal of Linguistics/Revue Canadienne de Linguistique* 62 (2): 314–334. https://doi.org/10.1017/cnj.2017.6.

R. Sasamoto, *Onomatopoeia and Relevance*, Palgrave Studies in Sound,
https://doi.org/10.1007/978-3-030-26318-8

Akita, Kimi, and Mark Dingemanse. 2019. Ideophones (Mimetics, Expressives). In *Oxford Research Encyclopedia of Linguistics*, ed. Mark Aronoff. Oxford: Oxford University Press.

Akita, Kimi, and Natsuko Tsujimura. 2016. Mimetics. In *Handbook of Japanese Lexicon and Word Formation*, ed. Taro Kageyama and Hideki Kishimoto, vol. 3, 133–160. Berlin and Boston: De Gruyter Mouton. https://doi.org/10.1515/9781614512097-008.

Akiyama, Madoka. 2002. "Syoku" o Meguru Hyoogen- Resipi No Bunsyoo/Buntai Ni Tuite [Expressions Related to "Food": On the Discourse/Style of Recipes]. *Kokugakuin Daigaku Kokugo Kenkyuukai "Kokugo Kenkyuu" [Kokugakuin University Japanese Research Society "Inquiries into the Japanese Language"]* 65: 34–46.

———. 2003. Syokkan o Arawasu Onomatope Ni Tuite [On the Onomatopoeia of the Texture]. *Kokugakuin Daigaku Kokugo Kenkyuukai "Kokugo Kenkyuu" [Kokugakuin University Japanese Research Society "Inquiries into the Japanese Language"]* 66: 12–27.

Amafuwa Girlie Wave. Hair & Beauty Miq no hea katarogu [Sweet and airly, girlie curly hair, Hair-style catalogue by Hair & Beauty miq]. n.d. Hotpepper Beauty. http://beauty.hotpepper.jp/slnH000231469/style/L003653319.html?vos=cphpbsoccap0111013001.

Ameka, Felix. 1992. Interjections: The Universal yet Neglected Part of Speech. *Journal of Pragmatics* 18 (2): 101–118. https://doi.org/10.1016/0378-2166(92)90048-G.

Aoyama, Tomoko. 2003. The Cooking Man in Modern Japanese Literature. In *Asian Masculinities: the Meaning and Practice of Manhood in China and Japan*, ed. K. Louie and M. Low, 155–176. London: Routledge Curzon.

Asaga, Chisato, Yusuf Mukarramah, and Chiemi Watanabe. 2008. ONOMATOPEDIA: Onomatopoeia Online Example Dictionary System Extracted from Data on the Web. In *Proceedings of the 10th Asia-Pacific Web Conference on Progress in WWW Research and Development*, 601–612. APWeb'08. Berlin and Heidelberg: Springer-Verlag. http://dl.acm.org/citation.cfm?id=1791734.1791805.

Asano, Tsuruko, ed. 1978. *Giongo-Gitaigo Jiten [The Phonomime Phenomime Dictionary]*. Tokyo: Kadokawa Shuppan.

Assaneo, Maria Florencia, Juan Ignacio Nichols, and Marcos Alberto Trevisan. 2011. The Anatomy of Onomatopoeia. *PLoS ONE* 6(12): e28317. https://journals.plos.org/plosone/article?id=10.1371/journal.pone.0028317.

Atoda, Toshiko, and Kazuko Hoshino. 1995. *Giongo/Gitaigo Tukaikata Jiten [Usage Guide to Japanese Onomatopoeias]*. Tokyo: Sotakusha.

Baker, Mona. 1992. *In Other Words: A Coursebook on Translation*. London: Routledge.

Bartashova, Olga A., and Anton E. Sichinskiy. 2014. Japanese–English Onomatopoeic and Mimetic Parallels: The Problem of Translatability. *Journal of Siberian Federal University. Humanities & Social Sciences* 2: 222–229.

Beeman, William O. 2014. Negotiating a Passage to the Meal in Four Cultures. In *Pragmatics & Beyond New Series*, ed. Polly E. Szatrowski, vol. 238, 31–52. Amsterdam: John Benjamins Publishing Company. https://doi.org/10.1075/pbns.238.02bee.

Berkeley Women and Language Group (University of California Berkeley). 1994. *Cultural Performances: Proceedings of the Third Berkeley Women and Language Conference, April 8, 9, and 10, 1994*. Berkeley, CA: Berkeley Women and Language Group, University of California.

Berndt, Jaqueline. 2011. Manga as "Media Geijutsu". *The Locality and Universality of "Media Geijutsu": Beyond "Cool Japan*. ICOMAG 2011.

Bianchi, Claudia, ed. 2004. *The Semantics/Pragmatics Distinction*. Chicago: CSLI Publications https://www.press.uchicago.edu/ucp/books/book/distributed/S/bo3627612.html.

Blakemore, Diane. 2002. *Relevance and Linguistic Meaning: The Semantics and Pragmatics of Discourse Markers*. Cambridge and New York: Cambridge University Press.

———. 2008. Apposition and Affective Communication. *Language and Literature* 17 (1): 37–57. https://doi.org/10.1177/0963947007085054.

———. 2011. On the Descriptive Ineffability of Expressive Meaning. *Journal of Pragmatics* 43 (14): 3537–3550. https://doi.org/10.1016/j.pragma.2011.08.003.

———. 2015. Slurs and Expletives: A Case against a General Account of Expressive Meaning. *Language Sciences* 52: 22–35.

Blathwayt, Benedict. 2005. *Little Red Train*. London: Random House Children's Books.

Bloom, Lynn Z. 2008. Consuming Prose: The Delectable Rhetoric of Food Writing. *College English* 70 (4): 346–362.

Bourdieu, Pierre. 1984. *Distinction—Pierre Bourdieu | Harvard University Press*. Cambridge, MA: Harvard University Press http://www.hup.harvard.edu/catalog.php?isbn=9780674212770.

Bradner, Liesl. 2009. Origins of Manga and Anime. *The Los Angeles Times*, November 27.

Brown, Penelope. 2011. Color Me Bitter: Crossmodal Compounding in Tzeltal Perception Words. *The Senses and Society* 6 (1): 106–116. https://doi.org/10.2752/174589311X12893982233957.

Bucholtz, Mary, A.C. Liang, and Laurel A. Sutton. 1999. *Reinventing Identities: The Gendered Self in Discourse*. Oxford: Oxford University Press.

Burapi & Jolie Sorotte Rainichi. Atsuatsu Kishakaiken [Brad Pitt and Angelina Jolie Visit Japan Together. Loved-up Press Conference]. 2005. Eiga.Conm. November 29. https://eiga.com/news/20051129/1/.

Burdenski, Matthew James. 2006. *Language Socialization of Two-Year Old Children in Kansai, Japan: The Family and Beyond*. Ann Arbor, MI: UCLA.

Carston, Robyn. 1997. Enrichment and Loosening: Complementary Processes in Deriving the Proposition Expressed? In *Pragmatik: Implikaturen und Sprechakte*, ed. Eckard Rolf, vol. 8, 103–127. Linguistische Berichte. Wiesbaden: VS Verlag für Sozialwissenschaften. https://doi.org/10.1007/978-3-663-11116-0_7.

———. 2002. *Thoughts and Utterances: The Pragmatics of Explicit Communication*. Oxford: Blackwell http://discovery.ucl.ac.uk/12106/.

———. 2010. Lexical Pragmatics, Ad Hoc Concepts and Metaphor: A Relevance Theory Perspective. *Italian Journal of Linguistics* 22: 153–180.

Casas-Tost, Helena. 2014. Translating Onomatopoeia from Chinese into Spanish: A Corpus-Based Analysis. *Perspectives* 22 (1): 39–55. https://doi.org/10.1080/0907676X.2012.712144.

Chiaro, Delia. 2010. *Translation, Humour and the Media: Translation and Humour*. London: Continuum.

Childs, G.Tucker. 1995. African Ideophones. In *Sound Symbolism*, ed. Leanne Hinton, Johanna Nichols, and John Ohala, 178–204. Cambridge and New York: Columbia University Press. https://doi.org/10.1017/CBO9780511751806.013.

Clancy, Patricia M. 1986. The Acquisition of Communicative Style in Japanese. In *Language Socialization Across Cultures*, ed. Bambi B. Schieffelin and Elinor Ochs, 213–250. New York: Cambridge University Press.

Clark, Billy. 2013. *Relevance Theory*. Cambridge: Cambridge University Press. https://doi.org/10.1017/CBO9781139034104.

Cloake, Felicity. 2016. How to Make the Perfect Cacio e Pepe. *The Guardian*, November 3, sec. Food. https://www.theguardian.com/lifeandstyle/wordofmouth/2016/nov/03/how-to-make-the-perfect-cacio-e-pepe.

Cohn, Neil. 2010. Japanese Visual Language: The Structure of Manga. In *Manga: An Anthology of Global and Cultural Perspectives*, ed. Toni Johnson-Woords, 187–203. https://research.tilburguniversity.edu/en/publications/japanese-visual-language-the-structure-of-manga.

———. 2014a. The Architecture of Visual Narrative Comprehension: The Interaction of Narrative Structure and Page Layout in Understanding

Comics. *Frontiers in Psychology* 5: 680. https://doi.org/10.3389/fpsyg.2014.00680.
———. 2014b. Visual Narrative Structure. *Cognitive Science* 37 (3): 413–452.
Cook, Haruko Minegishi. 2006. Joint Construction of Folk Beliefs by JFL Learners and Japanese Host Families. In *Language Learners in Study Abroad Contexts*, ed. Margaret A. DuFon and Eton E. Churchill, 120–150. Bristol: Multilingual Matters. https://doi.org/10.21832/9781853598531-009.
———. 2011. Language Socialization and Stance-Taking Practices. In *The Handbook of Language Socialization*, ed. Alessandro Duranti, Elinor Ochs, and Bambi B. Schieffelin, 296–321. Malden, MA: Blackwell. https://doi.org/10.1002/9781444342901.ch13.
Cooper-chen, Anne M. 2010. *Cartoon Cultures: The Globalization of Japanese Popular Media*. New York: Peter Lang Pub Inc.
Cotter, Colleen. 1997. Claiming a Piece of the Pie: How the Language of Recipes Defines Community. In *Recipes for Reading: Community Cookbooks, Stories, Histories*, ed. Anne Bower, 51–71. Amherst, MA: University of Amherst Press.
Counihan, Carole M. 2000. Fantasy Food: Gender and Food Symbolism in Preschool Children's Madeup Stories. In *The Anthropology of Food and Body: Gender, Meaning and Power*, ed. Carole M. Counihan, 129–155. Longon: Routledge.
———. 2018. Fantasy Food: Gender and Food Symbolism in Preschool Children's Madeup Stories. In *The Anthropology of Food and Body: Gender, Meaning and Power*, ed. Carole M. Counihan, 129–155. London: Routledge. https://doi.org/10.4324/9781315656540.
Cytowic, Richard E. 1997. Synesthesia: Phenomenology and Neuropsychology—A Review of Current Knowledge. In *Synaesthesia: Classic and Contemporary Readings*, ed. John E. Harrison and Simon Baron-Cohen, 17–42. Oxford: Blackwell.
Cytowic, Richard E., and Frank B. Wood. 1982. Synesthesia: II. Psychophysical Relations in the Synesthesia of Geometrically Shaped Taste and Colored Hearing. *Brain and Cognition* 1 (1): 36–49. https://doi.org/10.1016/0278-2626(82)90005-7.
Dailey, Audrey, Colin Martindale, and Jonathan Borkum. 1997. Creativity, Synesthesia and Physiognomic Perception. *Creativity Research Journal* 10 (1): 1–8. https://doi.org/10.1207/s15326934crj1001_1.
Dawson, Carl. 1990. *November 1948: A Memoir*. Charlottesville: University of Virginia Press.

Day, Sean. 2005. Some Demographic and Socio-Cultural Aspects of Synesthesia. In *Synesthesia: Perspectives From Cognitive Neuroscience*, ed. Lynn C. Robertson and Noam Sagiv, 11–33. Oxford: Oxford University Press.

———. 2009. Types-of-Syn. http://www.daysyn.com/Types-of-Syn.html.

DeVault, Marjorie L. 1994. *Feeding the Family*. Chicago: University of Chicago Press. https://www.press.uchicago.edu/ucp/books/book/chicago/F/bo3684531.html.

Diffloth, Gérard. 1972. Notes on Expressive Meaning. *Chicago Linguistic Society* 8: 440–447.

Dimitriu, Rodica. 2004. Omission in Translation. *Perspectives* 12 (3): 163–175. https://doi.org/10.1080/0907676X.2004.9961499.

Dingemanse, Mark. 2011a. Ezra Pound among the Mawu: Ideophones and Iconicity in Siwu. In *Semblance and Signification*, ed. Pascal Michelucci, Olga Fischer, and Christina Ljungberg, 39–54. Amsterdam: John Benjamins Publishing. https://www.mpi.nl/publications/446081.

———. 2011b. *The Meaning and Use of Ideophones in Siwu—Max Planck Institute for Psycholinguistics*. Nijmegen: Radboud University. https://www.mpi.nl/publications/1005623.

———. 2011c. Ideophones and the Aesthetics of Everyday Language in a West-African Society. *The Senses and Society* 6 (1): 77–85. https://doi.org/10.2752/174589311X12893982233830.

———. 2012. Advances in the Cross-linguistic Study of Ideophones. *Language and Linguistics Compass* 6 (10): 654–672. https://doi.org/10.1002/lnc3.361.

Dingemanse, Mark, and Kimi Akita. 2016. An Inverse Relation Between Expressiveness and Grammatical Integration: On the Morphosyntactic Typology of Ideophones, with Special Reference to Japanese. *Journal of Linguistics* 53 (3): 501–532. https://doi.org/10.1017/S002222671600030X.

Doke, Clement M. 1935. *Bantu Linguistic Terminology*. London: Longmans. https://www.cambridge.org/core/journals/africa/article/bantu-linguistic-terminology-by-c-m-doke-london-longmans-green-co1935-pp-v-237-6s-net/62F6964DDFCA3941CE2E2E241536FFF5.

Domino, George. 1989. Synesthesia and Creativity in Fine Arts Students: An Empirical Look. *Creativity Research Journal* 2 (1–2): 17–29. https://doi.org/10.1080/10400418909534297.

Dubovičienė, Tatjana, and Pavel Skorupa. 2014. The Analysis of Some Stylistic Features of English Advertising Slogans. *Žmogus Ir Žodis* 16 (3): 61–75. https://doi.org/10.15823/zz.2014.013.

DuFon, Margaret A. 2006. The Socialization of Taste during Study Abroad in Indonesia. In *Language Learners in Study Abroad Contexts*, ed. Margaret

A. DuFon and Eton E. Churchill, 120–150. Clevedon: Multilingual Matters. https://doi.org/10.21832/9781853598531-008.

DuFon, Margaret A., and Eton E. Churchill. 2006. *Language Learners in Study Abroad Contexts*. Bristol: Multilingual Matters. https://doi.org/10.21832/9781853598531.

Duranti, Alessandro, Elinor Ochs, and Bambi B. Schieffelin, eds. 2011. *The Handbook of Language Socialization*. Malden, MA: Blackwell. https://doi.org/10.1002/9781444342901.ch13.

Earhart, H. Byron. 2011. Mount Fuji: Shield of War, Badge of Peace. *The Asia-Pacific Journal—Japan Focus*. September 5. https://apjjf.org/2011/9/20/H.-Byron-Earhart/3528/article.html.

Edström, Bert. 1989. Japanese Onomatopoetic Words: A Research Note. *Orientaliska Studier* 65: 35–52.

Eisner, Will. 2008. *Comics and Sequential Art: Principles and Practices from the Legendary Cartoonist*. 1st ed. New York: W. W. Norton & Company.

Empire, Kitty. 2014. Trouble in Paradise Review—La Roux Is Back, on Her Own. *The Observer*, July 20, sec. Music. https://www.theguardian.com/music/2014/jul/20/trouble-in-paradise-la-roux-back-synth-pop-review.

Enfield, Nicholas J. 2011. Taste in Two Tongues: A Southeast Asian Study of Semantic Convergence. *The Senses and Society* 6 (1): 30–37. https://doi.org/10.2752/174589311X12893982233632.

Evans, Jessica, and Stuart Hall. 1999. *Visual Culture: The Reader*. London and Thousand Oaks: Sage Publications in Association with the Open University.

Fillmore, Charles. 1982. Frame Semantics. In *Linguistics in the Morning Calm*, ed. The Linguistics Society of Korea, 111–137. Seoul: Hanshin. https://www.coursehero.com/file/29281284/Fillmore-1982-Frame-Semanticspdf/.

———. 1985. Frames and the Semantics of Understanding. *Quaderni Di Semantica* 6: 222–254.

Fisher, Mary Frances Kennedy. 1969. The Anatomy of a Recipe. In *With Bold Knife and Fork*, 1–14. New York: Counterpoint.

Flyxe, Martin. 2002. Translation of Japanese Onomatopoeia Into Swedish (with Focus on Lexicalization). *Africa & Asia* 2: 54–73.

Fodor, J., J. Fodor, and M. Garrett. 1975. The Psychological Unreality of Semantic Representations. *Linguistic Enquiry* 6: 515–531.

Fujino, Yoshitaka. 2009. Tensai Asuriito No Oto [The Sound of Genius Athletes]. *Watashi No Onomatope Sutairu [My Onomatopoeia Style]*. http://happy-vocalism.cocolog-nifty.com/blog/cat41995244/.

Fukuma, Yasuko. 1993. Manga-Ni Miru Giongo/Gitaigo-No Tokuisei-Ni Tuite [On the Peculiarity of Mimetics in Cartoons]. *Kyuusyuu Daigaku Ryuugakusei-*

Kyooiku-Sentaa Kiyoo [Bulletin of the International Student Center of Kyushu University] 5: 185–196.

Gallace, Alberto, Erica Boschin, and Charles Spence. 2011. On the Taste of "Bouba" and "Kiki": An Exploration of Word–Food Associations in Neurologically Normal Participants. *Cognitive Neuroscience* 2 (1): 34–46. https://doi.org/10.1080/17588928.2010.516820.

Garrett, Paul B., and Patricia Baquedano-López. 2002. Language Socialization: Reproduction and Continuity, Transformation and Change. *Annual Review of Anthropology* 31 (1): 339–361. https://doi.org/10.1146/annurev.anthro.31.040402.085352.

Gherasim, Toader. 2014. The Concept of a Brand. *Economy Transdisciplinarity Cognition* 17: 8–12.

Goddard, Cliff. 2011. *Semantic Analysis: A Practical Introduction*. 2nd ed. Oxford Textbooks in Linguistics. Oxford and New York: Oxford University Press.

Goffman, Erving. 1981. *Forms of Talk*. Oxford: Blackwell. http://www.upenn.edu/pennpress/book/715.html.

Goodwin, Charles. 1981. *Conversational Organization: Interaction between Speakers and Hearers*. New York: Academic Press.

———. 1986. Between and within: Alternative Sequential Treatments of Continuers and Assessments. *Human Studies* 9 (2): 205–217. https://doi.org/10.1007/BF00148127.

———. 2000. Action and Embodiment within Situated Human Interaction. *Journal of Pragmatics* 32: 1489–1522.

———. 2003. Pointing as Situated Practice. In *Pointing: Where Language, Culture, and Cognition Meet*, ed. Sotaro Kita, 217–241. Mahwah, NJ: Lawrence Erlbaum Associates.

———. 2007. Environmentally Coupled Gestures. In *Gesture and the Dynamic Dimension of Language*, ed. Susan D. Duncan, Justine Cassell, and Elena T. Levy, 195–212. Amsterdam: John Benjamins Publishing. https://benjamins.com/catalog/gs.1.18goo.

Goodwin, Charles, and Marjorie Harness Goodwin. 1987. Concurrent Operations on Talk: Notes on the Interactive Organization of Assessments. *IPRA Papers in Pragmatics* 1 (1): 1–54.

Goodwin, Marjorie Harness, and Charles Goodwin. 2001. Emotion within Situated Activity. In *Linguistic Anthropology: A Reader*, ed. Alessandro Duranti, 239–257. Oxford: Blackwell.

Goodwin, Marjorie, Asta Cekaite, and Charles Goodwin. 2012. Emotion as Stance. In *Emotion in Interaction*, ed. Anssi Perakyla and Marja-Leena

Sorjonen, 16–40. Oxford: Oxford University Press. https://doi.org/10.1093/acprof:oso/9780199730735.003.0002.

Gravett, Paul. 2004. *Manga: Sixty Years of Japanese Comics*. 1st ed. New York: Harper Design.

Grice, H. Paul. 1957. Meaning. *Philosophical Review* 66 (3): 377–388.

———. 1989. *Studies in the Way of Words*. Cambridge, MA: Harvard University Press. http://www.hup.harvard.edu/catalog.php?isbn=9780674852716.

Griffiths, Owen. 2007. Militarizing Japan: Patriotism, Profit, and Children's Print Media, 1894–1925. *The Asia-Pacific Journal—Japan Focus*, March 9. https://apjjf.org/-Owen-Griffiths/2528/article.html.

Guynes, Sean A. 2014. Four-Color Sound: A Peircean Semiotics of Comic Book Onomatopoeia. *The Public Journal of Semiotics* 6: 58–72.

Halliday, G. 2015. *Homicide in High Heels: High Heels Mysteries Book #8*. High Heels Mysteries. Gemma Halliday Publishing. https://books.google.ie/books?id=Ae5VBgAAQBAJ.

Hamano, Shoko. 1998. *The Sound-Symbolic System of Japanese*. Stanford, CA: CSLI Publications. http://ci.nii.ac.jp/ncid/BA38075516.

Harada, Aya. 2012. Resipi No Buntai Ni Kansuru Kenkyuu- Zassi "Huzin No Tomo" No Ryoori Kizi o Taisyoo Ni- [Research on the Style Used in Recipes: Recipe Articles in the Magazine Huzin No Tomo "Women's Friend"]. *Otyanomizu Zyosi Daigaku Kokugo Kokubun Gakkai "Kokubun" [Ochanomizu University Kokugo Kokubun Society "National Literature"]* 17: 51–61.

Harashima, Jun, Michiaki Ariga, Kenta Murata, and Masayuki Ioki. 2016. A Large-Scale Recipe and Meal Data Collection as Infrastructure for Food Research. In *Proceedings of the 10th International Conference on Language Resources and Evaluation*, 2455–2459. https://www.semanticscholar.org/paper/A-Large-scale-Recipe-and-Meal-Data-Collection-as-Harashima-Ariga/8660210d71d56c480bd568f49fcf5243ec80a84f.

Haryu, Etsuko, and Lihua Zhao. 2007. Yuuseion-to Museion-o Daisyoo-Ni Taioo-Zukeru Kankaku-No Kigen: Giongo-Rikai-No Nit-/Tyuu-Hikaku [Understanding the Symbolic Values of Japanese Onomatopoeia: Comparison of Japanese and Chinese Speakers]. *Japanese Psychological Research* 78: 424–432.

Hayakawa, Fumiyo. 2006. *Syokkan No Onomatope [Onomatopoeia for Food Texture]*. Tokyo: Sanseido.

Hayakawa, Fumiyo, Yufuko Iwamasa, Keiko Hatae, and Atsuko Shimada. 1999. Collection of Onomatopoeic Terms for Describing Food Properties. *Nihon Kasei Gakkaish [Journal of Home Economics Japan]* 50 (5): 481–491.

Hayakawa, Fumiyo, Shun-Sheng Chen, Xi-Chang Wang, Zai-Qui Li, Masayoshi Saito, Yasumasa Baba, and Masahito Yokoyama. 2004. Collection and Classification of Chinese Texture Terms: Studies on Chinese Texture Terms. *Nippon Syokuhin Kagaku Koogaku Kaisi [Food Science and Technology Research]* 51 (3): 131–141.

Hayakawa, Fumiyo, Kana Ioku, Sayuri Akuzawa, Masayoshi Saito, Katsuyoshi Nishinari, Yoshimasa Yamano, and Kaoru Kohyama. 2005. Collection of Japanese Texture Terms: Studies on Japanese Texture Terms. *Nippon Shokuhin Kagaku Kogaku Kaishi [Japanese Society for Food Science and Technology]* 52 (8): 337–346.

Hayakawa, Fumiyo, Kana Ioku, Sayuri Akuzawa, Chie Yoneda, Yukari Kazami, Katsuyoshi Nishinari, Yasumasa Baba, and Kaoru Kohyama. 2006. Shitsumon Shihoo Ni Yoru Shoohisha No Tekusuchaa Goi Choosa [Research Survey of Japanese Consumers on Texture Vocabulary]. *Nippon Shokuhin Kagaku Kogaku Kaishi [Journal of the Japanese Society for Food Science and Technology]* 53 (6): 327–336.

Hayakawa, Fumiyo, Sayuri Akuzawa, Chie Yoneda, Yukari Kazami, Katsuyoshi Nishinari, Yoshihiro Nakamura, Yasumasa Baba, and Kaoru Kohyama. 2007. Seebetu/Nenreebetu/Tiikibetu Ni Mita Nihongo Tekusutyaa Yoogo No Ninti Zyookyoo [Recognition of Japanese Texture Descriptive Terms According to Gender, Age and Region]. *Nippon Syokuhin Kagaku Koogaku Kaisi [Food Science and Technology Research]* 54 (11): 488–502.

Hayase, Mitsuaki. 1978. Characteristics of Japanese Onomatopoeia Observed Through the English Translation. *Kanssai Gaikokugodaigaku Kenkyu Ronshu* 28: 117–127.

Heer, Jeet, and Kent Worcester. 2009. *A Comics Studies Reader*. Jackson, MS: University Press of Mississippi.

Hines, Caitlin. 1994. "Let Me Call You Sweetheart": The Woman as Dessert Metaphor. In *Cultural Performances: Proceedings of the Third Berkeley Women and Language Conference*, ed. Mary Bucholtz, A.C. Liang, Laurel A. Sutton, and Caitlin Hines, 295–303. Berkeley, CA: Berkeley Women and Language Group.

———. 1999. Rebaking the Pie: The Woman as Dessert Metaphor. In *Reinventing Identities: The Gendered Self in Discourse*, ed. Mary Bucholtz, A.C. Liang, and Laurel A. Sutton, 145–162. Oxford: Oxford University Press.

Hinton, Leanne, Johanna Nichols, and John Ohala, eds. 1995a. *Sound Symbolism*. Cambridge: Cambridge University Press. https://doi.org/10.1017/CBO9780511751806.001.

———. 1995b. Introduction: Sound-Symbolic Processes. In *Sound Symbolism*, ed. Leanne Hinton, Johanna Nichols, and John Ohala, 1–12. Cambridge: Cambridge University Press. https://doi.org/10.1017/CBO9780511751806.001.

Hirata-Mogi, Sachiko, Jun Ukita, and Shinichi Kita. 2011. Implicit Phonetic Symbolism in Voicing of Consonants and Visual Lightness Using Gerner's Speeded Classification Task. *Perceptual and Motor Skills* 113 (December): 929–940. https://doi.org/10.2466/15.21.28.PMS.113.6.929-940.

'Home: Oxford English Dictionary'. n.d. https://www-oed-com.dcu.idm.oclc.org/. Accessed 12 May 2019.

Horsey, Richard. 1998. *Null Arguments in English Registers*. Thesis, La Trobe University, Bundoora.

Hoshino, Yuko. 2015. Gurume Kiji Ni Okeru Onomatope [Onomatopoeia in Gourmet Articles]. *Japanese Language Education in Europe* 19: 90–94.

Hosoma, Hiromichi. 2017. *Japanese Language Education in Europe*. Belfast.

House, Jill. 2006. Constructing a Context with Intonation. *Journal of Pragmatics* 38 (10): 1542–1558.

Hyuga, Shigeo. 1986. Manga No Giongo/Gitaigo (2) [Onomatopoeia in Manga]. *Nihongogaku* 15 (8): 98–108.

Ibarretxe-Antunano, Iraide. 2006. *Sound Symbolism and Motion in Basque*. Munich: LINCOM.

Imai, Mutsumi, and Sotaro Kita. 2014. The Sound Symbolism Bootstrapping Hypothesis for Language Acquisition and Language Evolution. *Philosophical Transactions of the Royal Society B: Biological Sciences* 369 (September): 1–13. https://doi.org/10.1098/rstb.2013.0298.

Imai, Mutsumi, Sotaro Kita, Miho Nagumo, and Hiroyuki Okada. 2008. Sound Symbolism Facilitates Early Verb Learning. *Cognition* 109 (1): 54–65. https://doi.org/10.1016/j.cognition.2008.07.015.

Inose, Hiroko. 2008. Translating Japanese onomatopoeia and mimetic words. In *Translation Research Projects*, ed. Anthony Pym and Alexander Perekrestenko, vol. 1, 97–116. Tarragona: Intercultural Studies Group.

Inoue, Kazuko. 2013. Onomatope No Tagisei to Sōzōsei [Polysemy and Creativity of Onomatopoeia]. In *OnomatopeKenkyū No Shatei: ChikazukuOto to Imi [Sound Symbolism and Mimetics: Rethinking the Relationship Between Sound and Meaning in Language]*, ed. Kazuko Shinohara and Ryoko Uno, 203–216. Tokyo: Hitsuji Shobo.

Itoi, Michihiro. 1989. Buntai to Shite Mita "Manga No Kotoba" [Language of Manga as a Style]. *Nihongogaku* 8–9: 66–73.

Ivanova, Gergana. 2006. Sound-Symbolic Approach to Japanese Mimetic Words. *Toronto Working Papers in Linguistics* 26: 103–114.

Iwasaki, Noriko, Peter Sells, and Kimi Akita. 2016. *The Grammar of Japanese Mimetics: Perspectives from Structure, Acquisition, and Translation*. London: Routledge.

Kadooka, Kenichi. 2005. On the Degree of Lexicalization in English Onomatopoeia from a Historical Perspective. *The Ryukoku Journal of Humanities and Sciences* 27 (1): 1–13.

Kageyama, Taro, and Hideki Kishimoto. 2016. *Handbook of Japanese Lexicon and Word Formation*. Berlin and Boston: De Gruyter Mouton. https://doi.org/10.1515/9781614512097.

Kagitani, Tatsuki, Mao Goto, Junji Watanbe, and Maki Sakamoto. 2014. Sound Symbolic Relationship between Onomatopoeia and Emotional Evaluations in Taste. *Proceedings of the Annual Meeting of the Cognitive Science Society* 36: 2871–2876.

Kaindl, Klaus. 2010. Comics in Translation. In *Handbook of Translation Studies*, ed. Yves Gambier and Luc van Doorslaer, 36–44. Amsterdam: John Benjamins Publishing. https://benjamins.com/catalog/hts.1.comi1.

Kakehi, H., I. Tamori, and L. Schourup. 1996. *Dictionary of Iconic Expressions in Japanese*. Berlin: Mouton de Gruyte.

Kantartzis, Katerina, Mutsumi Imai, and Sotaro Kita. 2011. Japanese Sound-Symbolism Facilitates Word Learning in English-Speaking Children. *Cognitive Science* 35 (3): 575–586.

Kaplan, David. 1997. *What Is Meaning? Explorations in the Theory of Meaning as Use*.

Karatsu, Mariko. 2004. Verbal and Nonverbal Negotiation in Japanese Storytelling. In *Hidden and Open Conflict in Japanese Conversational Interaction*, ed. Polly Szatrowski, 125–161. Tokyo: Kuroshio Shuppan.

———. 2012. *Conversational Storytelling among Japanese Women*. Amsterdam: John Benjamins Publishing. https://benjamins.com/catalog/sin.16.

Kato, Ayumi, Yusuke Fukazawa, and Taketoshi Mori. 2012a. Discovering Pattern of Onomatopoeia Used in Food Reviews. In *The 6th International Conference on Soft Computing and Intelligent Systems, and The 13th International Symposium on Advanced Intelligence Systems*, 502–507. Kobe, Japan. https://doi.org/10.1109/SCIS-ISIS.2012.6505214.

Kato, Ayumi, Yusuke Fukazawa, Tomomasa Sato, and Taketoshi Mori. 2012b. Extraction of Onomatopoeia Used for Foods from Food Reviews and Its Application to Restaurant Search. In Lyon, France. https://doi.org/10.1145/2187980.2188192.

Keene, Carolyn. 2003. *Intrigue at the Grand Opera (Nancy Drew, #171)*. New York: Alladin Paperbacks.

Kenji, Miyazawa, Sarah M. Strong, and Karen Colligan-Taylor. 2002. *Masterworks of Miyazawa Kenji – Poems and Fairy Tales*. Tokyo: International Foundation for the Promotion of Languages and Culture.

Kern, Adam L. 2006. *Manga from the Floating World: Comicbook Culture and the Kibyōshi of Edo Japan*. Vol. 279. 1st ed. Cambridge, MA: Harvard University Asia Center. https://doi.org/10.2307/j.ctt1tg5pwh.

Kindaichi, Haruhiko. 1978. Giongo/Gitaigo Gaisetu [Overview of Mimetics]. In *Giongo/Gitaigo Ziten [A Dictionary of Mimetics]*, ed. Tsuruko Asano, 3–25. Tokyo: Kadokawa Shuppan.

Kinsella, Sharon. 1999. Pro-Establishment Manga: Pop-Culture and the Balance of Power in Japan. *Media, Culture & Society* 21 (4): 567–572. https://doi.org/10.1177/016344399021004007.

Kita, Sotaro. 1997. Two-Dimensional Semantic Analysis of Japanese Mimetics. *Linguistics* 35 (January): 379–416. https://doi.org/10.1515/ling.1997.35.2.379.

———. 2001. Semantic Schism and Interpretive Integration in Japanese Sentences with a Mimetic: A Reply to Tsujimura. *Linguistics* 39 (January): 419–436. https://doi.org/10.1515/ling.2001.017.

———. 2013. Giongo Gitaigo to Jihatsuteki Miburi. In *Onomatope Kenkyu No Shatei: Chikazuku Oto to Imi*, ed. Kazuko Shinohara and Ryoko Uno, 79–84. Tokyo: Hitsuji Shobo.

Kitamura, H. 1987. Television Is a Fiction and Viewers Are Observers. *Studies of Broadcasting* 23: 141–153.

Köhler, Wolfgang. 1929. *Gestalt Psychology*. New York: H. Liveright.

Koike, Chisato. 2001. An Analysis of Shifts in Participation Roles in Japanese Storytelling in Terms of Prosody, Gaze and Body Movements. *Proceedings of the Twentyseventh Annual Meeting of the Berkeley Linguistics Society* (Vol. 27, pp. 381–392).

———. 2009. *Interaction in Storytelling in Japanese Conversations: An Analysis of Story Recipients' Questions*. Doctoral dissertation, University of California at Los Angeles, Los Angeles, CA, USA.

———. 2010. Ellipsis and Action in a Japanese Joint Storytelling Series: Gaze, Pointing, and Context. In *Storytelling Across Japanese Conversational Genre*, ed. P. Szatrowski, 61–111. Amsterdam: John Benjamins.

Koo, Yeon-Hwa. 2009. An Advertisement and Food: Vocabulary of a Television Commercial in Japan and Korean. *The 3rd Consortium on Global Perspectives in Japanese Studies*.

Kubo, Atsuko. 1995. Miyazawa Kenji No Onomatope No Sekai [The World of Onomatopoeia in Miyazawa Kenji's Work]. *Kobe Kaisei Review* 130 (34): 17–30.

Kubozono, Haruo. 2015. *Handbook of Japanese Phonetics and Phonology*. Berlin and Boston: De Gruyter Mouton. https://doi.org/10.1515/9781614511984.

Kulick, Don, and Bambi Schieffelin. 2004. Language Socialization. In *A Companion to Linguistic Anthropology*, ed. Alessandro Duranti, 349–368. Oxford: Blackwell.

Kuroshima, Satomi. 2010. Another Look at the Service Encounter: Progressivity, Intersubjectivity, and Trust in a Japanese Sushi Restaurant. *Journal of Pragmatics* 42 (3): 856–869. https://doi.org/10.1016/j.pragma.2009.08.009.

———. 2014. The Structural Organization of Ordering and Serving Sushi. In *Language and Food: Verbal and Nonverbal Experiences*, ed. Polly Szatrowski, 54–75. Amsterdam: John Benjamins Publishing https://benjamins.com/catalog/pbns.238.03kur.

Kusamori, Shinichi. 1968. Sutoori Manga No Bunpo Ron. Giseigo Gitaigo No Ronri [Grammar of Story Manga: Logic of Onomatopoeia]. *Com.*

Kusumi, Takashi. 2005. Kokoro de Ajiwau—Mikaku Hyogen o Sasaeru Ninchi No Shikumi [Taste with Mind—The Cognitive Mechanism for Taste Expressions]. In *Aji Kotoba No Sekai [The World of Tasty Expressions]*, ed. Kenichi Seto, Takashi Kusumi, Shigeo Sawai, Tomoko Tsujimoto, Haruhiko Yamaguchi, and Shunsuke Koyama, 88–122. Tokyo: Kaimeisha.

Kwon, Nahyun. 2018. Iconicity Correlated with Vowel Harmony in Korean Ideophones. *Laboratory Phonology: Journal of the Association for Laboratory Phonology* 9 (1): 1. https://doi.org/10.5334/labphon.53.

Laing, Catherine E. 2014. A Phonological Analysis of Onomatopoeia in Early Word Production. *First Language* 34 (5): 387–405. https://doi.org/10.1177/0142723714550110.

———. 2017. A Perceptual Advantage for Onomatopoeia in Early Word Learning: Evidence from Eye-Tracking. *Journal of Experimental Child Psychology* 161: 32–45. https://doi.org/10.1016/j.jecp.2017.03.017.

Laing, Catherine. 2018. A Role for Onomatopoeia in Early Language: Evidence from Phonological Development. *Language and Cognition* 11 (2): 1–15.

Laing, Catherine E. 2019. Phonological Motivation for the Acquisition of Onomatopoeia: An Analysis of Early Words. *Language Learning and Development* 15 (2): 117–197.

Lakoff, Robin Tolmach. 2006. Identity à La Carte: You Are What You Eat. In *Discourse and Identity*, ed. Anna De Fina, Deborah Schiffrin, and Michael Bamberg, 142–165. Amsterdam: John Benjamins Publishing. https://doi.org/10.1017/CBO9780511584459.008.

Lawson, Nigella. 2001. Creme Brulée. *Nigella.Com*. http://www.nigella.com/recipes/creme-brulee.

———. 2012. Soba Noodles With Sesame Seeds. *Nigella.Com*. http://www.nigella.com/recipes/soba-noodles-with-sesame-seeds.

Lischer, Brian. 2015. 7 Popular Types of Brand Names. *Ignyte* (blog), October 19. http://www.ignytebrands.com/7-popular-types-of-brand-names/.

MacWilliams, Mark W. 2008. *Japanese Visual Culture: Explorations in the World of Manga and Anime*. London: Routledge. https://www.goodreads.com/work/best_book/3148567-japanese-visual-culture-explorations-in-the-world-of-manga-and-anime.

Magee, H. Melissa. 2009. Talking About Taste: How the Description of Food Means and Does. Swarthmore College, BA Thesis. http://www.swarthmore.edu/sites/default/files/assets/documents/linguistics/2009_Magee.pdf.

Majid, Asifa, and Stephen C. Levinson. 2011. The Senses in Language and Culture. *The Senses and Society* 6 (1): 5–18. https://doi.org/10.2752/174589311X12893982233551.

Mannion, Niamh. 2016. NINA Capri: Recipes from Italy's Amalfi Coast | Cookbook Review. *TheTaste.ie* (blog), July 4. http://thetaste.ie/wp/a-taste-of-la-dolce-vita-in-nina-capri-cookbook-review/.

Marechal, Beatrice. 2001. The Singular Stories of the Terashima Neighborhood: A Japanese Autobiographical Comic. *International Journal of Comic Art* 3 (2): 138–150.

Maria, Teresa. 2018. Day 10: About High Heel-Obsession and How to Find the Perfect Pair. *Outlandish* (blog), August 10. https://outlandishblog.com/about-high-heels/.

Marshall, Shelly. 2014. *Night on the Milky Way Railroad* (Kindle Edition). Los Gatos: Smashwords.

Martin, Samuel E. 1975. *A Reference Grammar of Japanese*, Yale Linguistics Series. Vol. 40. New Haven, CT and London: Yale University Press.

Matsuo, Shoko. 2014. *Studies of Japanese Expressions Describing Food*. Kyoto: Kyoto Prefectural University.

Matwick, Keri, and Kelsi Matwick. 2014. Storytelling and Synthetic Personalization in Television Cooking Shows. *Journal of Pragmatics* 71 (September): 151–159. https://doi.org/10.1016/j.pragma.2014.08.005.

Maurer, Daphne, Thanujeni Pathman, and Catherine J. Mondloch. 2006. The Shape of Boubas: Sound–Shape Correspondences in Toddlers and Adults. *Developmental Science* 9 (3): 316–322. https://doi.org/10.1111/j.1467-7687.2006.00495.x.

McCarthy, Helen. 2010. Ippei Okamoto: Notes Towards a List of Works. *Helen McCarthy: A Face Made for Radio* (blog), September 16. https://helenmccarthy.wordpress.com/2010/09/16/ippei-okamoto-notes-towards-a-list-of-works-2/.

McCawley, James D. 1968. *The Phonological Component of a Grammar of Japanese*. Hague and Paris: Mouton.

McCloud, Scott. 1993. *Understanding Comics: The Invisible Art*. Princeton, NJ: Kitchen Sink Press.

———. 2006. *Making Comics: Storytelling Secrets of Comics, Manga and Graphic Novels*. New York: Harper.

Meinard, Maruszka Eve Marie. 2015. Distinguishing Onomatopoeias from Interjections. *Journal of Pragmatics Complete* 76: 150–168. https://doi.org/10.1016/j.pragma.2014.11.011.

Mikami, Kyoko. 2004. Tagi Onomatopoeia No Imi/Yoho No Kijutsu to Shido No Kokoromi: "Gorogoro" "Batabata" o Rei to Shite. *Koide Kinen Nihongo Kyoiku Kenkyukai Ronbunshu* 12: 63–77.

Minashima, Hiroshi. 2004. Nihhongo No Onomatope [Onomatopoeia in Japanese]. *Fukui Daigaku Kyoiku Chiiki Kagakubu Kiyou* 60: 97–115.

Mithun, Marianne. 1982. The Synchronic and Diachronic Behavior of Plops, Squeaks, Croaks, Sighs, and Moans. *International Journal of American Linguistics* 48 (1): 49–58.

Miyazawa, Kenji. 1951. *Kaze no Matasaburo [Matasaburo of the Wind]*. Tokyo: Iwanami Shoten. https://www.aozora.gr.jp/cards/000081/files/462_15405.html.

Mizuta, Naomi. 2007. Animeeshon to Manga No Onomatope [Onomatopoeia in Animation and Manga]. *The Bulletin of Kurashiki University of Science and the Arts* 12: 197–204.

Molina, Lucia. 2006. *El Otoño Del Pingüino. Análisis Descriptivo de La Traducción de Los Culturemas*. Castelló de la Plana: Publicacions de la Universitat Jaume I.

Moon, Tom. 2014. First Listen: Adult Jazz, "Gist Is". *NPR.Org*, July 27. https://www.npr.org/2014/07/27/333716267/first-listen-adult-jazz-gist-is.

Moriyama, Keiko. 2007. Onomatope no Moto: Kan/Gan [The Foundation of Onomatopoeia: Kan/Gan]. In *Nihongo Onomatope Jiten [Japanese Onomatopoeia Dictionary]*, ed. Masahiro Ono, 57. Tokyo: Shogakukan Publishing.

Morooka, Tomonori. 2010. The Onomatopoeias in Manga. *Studies in Literature and Culture* 47: 15–24.

Mulvenna, Catherine, Edward Hubbard, Vilayanur Ramachandran, and Frank Pollick. 2004. The Relationship Between Synaesthesia and Creativity. In *Eleventh Annual Meeting of the Cognitive Neuroscience Society*. San Francisco.

Murakami, Haruki. 1999. *Suputoniku no Koibito [Sputnik Sweeaheart]*. Tokyo: Kodansha.

Muto, Ayaka. 2001. Mikaku Keiyoshi "Amai" to "Karai" No Tagi Kozo. [Polysemous Structure of Taste Adjectives Amai and Karai]. *Journal of Japanese Language Teaching* 110: 42–51.

———. 2003. Aji Kotoba No Giongo Gitaigo: Shoku o Onomatope [Onomatopoeia in Taste Expressions: Onomatopoeia for Food]. In *Kotoba Wa Aji o Koeru*, ed. Kenichi Seto, 241–300. Tokyo: Kaimeisha.

Nangumo. 2014. Shittori Amafuwa Rennyu-Pan [Moist and Sweet-Airy Milk Bread]. *CookPad*. https://cookpad.com/recipe/2684995.

Nash, Eric P. 2009. *Manga Kamishibai: The Art of Japanese Paper Theater*. New York: Harry N. Abrams.

Nasu, Akiko. 2015. The Phonological Lexicon and Mimetic Phonology. In *Handbook of Japanese Phonetics and Phonology*, ed. Haruo Kubozono, 253–288. Berlin: Mouton de Gruyter.

Natsume, Fusanosuke. 1997. *Manga No Bunpo: Komikkugaku No Mikata [Grammar of Manga: Perspectives on Comic Studies]*. Tokyo: Asahi Shimbunsha.

———. 2013. Manga Ni Okeru Onomatope [Onomatopoeia in Manga]. In *Onomatope Kenkyu No Shatei: Chikazuku Oto to Imi*, ed. Kazuko Shinohara and Ryoko Uno, 217–241. Tokyo: Hitsuji Shobo.

Newman, John, ed. 2009. *The Linguistics of Eating and Drinking*. Amsterdam: John Benjamins Publishing. https://benjamins.com/catalog/tsl.84.

Newman, Paul. 2001. Are Ideophones Really as Weird and Extra-Systematic as Linguists Make Them Out to Be? In *Ideophones*, ed. F.K.E. Voeltz and C. Kilian-Hatz, 251–258. Amsterdam: John Benjamins.

Noh, Eun-Ju. 2000. *Metarepresentation: A Relevance-Theory Approach*. Amsterdam: John Benjamins Publishing.

Norrick, Neal R. 2011. Conversational Recipe Telling. *Journal of Pragmatics* 43 (11): 2740–2761. https://doi.org/10.1016/j.pragma.2011.04.010.

Nuckolls, Janis B. 1996. *Sounds Like Life: Sound-Symbolic Grammar, Performance, and Cognition in Pastaza Quechua*. New York: Oxford University Press.

———. 2010. The Sound-Symbolic Expression of Animacy in Amazonian Ecuador. *Diversity* 2 (3): 353–369. https://doi.org/10.3390/d2030353.

Nygaard, Lynne C., Allison E. Cook, and Laura L. Namy. 2009. Sound to Meaning Correspondences Facilitate Word Learning. *Cognition* 112 (1): 181–186. https://doi.org/10.1016/j.cognition.2009.04.001.

O'Hagan, Minako. 2010. Japanese TV Entertainment: Framing Humour with Open Caption Telop. In *Translation, Humour and the Media: Translation and Humour*, ed. Delia Chiaro, vol. 2, 70–88. London: Continuum.

Ochs, E. 2002. Becoming a Speaker of Culture. In *Language Socialization and Language Acquisition: Ecological Perspectives*, ed. C. Kramsch, 99–120. New York: Continuum Press.

Ochs, Elinor, and Bambi B. Schieffelin. 1982. Language Acquisition and Socialization: Three Developmental Stories and Their Implications. Sociolinguistic Working Paper Number 105. In *Cultural Theory: Essays on Mind, Self and Emotion*, ed. Richard Schweder and Robert A. LeVine. New York: Cambridge University Press. https://eric.ed.gov/?id=ED252065.

Ochs, Elinor, and Merav Shohet. 2006. The Cultural Structuring of Mealtime Socialization. *New Directions for Child and Adolescent Development* 2006 (111): 35–49. https://doi.org/10.1002/cd.154.

Ochs, Elinor, Clotilde Pontecorvo, and Alessandra Fasulo. 1996. Socializing Taste. *Ethnos* 61 (1–2): 7–46. https://doi.org/10.1080/00141844.1996.9981526.

Ohori, Toshio. 2018. *Ninchi Gengogaku to Goyoron: Gengo Kigo No "Imi" [Cognitive Linguistics and Pragmatics: "Meaning" of Linguistic Symbols].* Tokyo: Kyorin University.

Oliver, Jamie. 2013. *Pork Stuffing | Pork Recipes | Jamie Oliver Recipes*. Jamie Oliver. https://www.jamieoliver.com/recipes/pork-recipes/stuffing/.

Ono, Masahiro. 2007. *Nihongo Onomatope Jiten [Japanese Onomatopoeia Dictionary]*. Tokyo: Shogakukan Publishing.

Onomatopoeia Sounds Like. n.d. https://www.stage32.com/sites/stage32.com/files/assets/screenplay/185490/25610_1536758815.pdf.

Oswalt, Robert L. 1995. Inanimate Imitatives in English. In *Sound Symbolism*, ed. Leanne Hinton, Johanna Nichols, and John Ohala, 293–306. Cambridge: Cambridge University Press. https://doi.org/10.1017/CBO9780511751806.020.

Özdamar, Melda. 2012. *The Visual Language of Cook Books*. Izmir: Izmir University of Economics.

Parault, Susan J., and Paula J. Schwanenflugel. 2006. Sound-Symbolism: A Piece in the Puzzle of Word Learning. *Journal of Psycholinguistic Research* 35 (4): 329–351.

Petersen, Robert S. 2009. The Acoustics of Manga. In *A Comics Studies Reader*, ed. Jeet Heer and Kent Worcester, 163–172. Jackson, MS: University Press of Mississippi. http://www.jstor.org/stable/j.ctt2tvd9s.22.

Pilkington, Adrian. 2000. *Poetic Effects*. Amsterdam: John Benjamins Publishing. https://benjamins.com/catalog/pbns.75.

Poe, Edgar Allan. 2001. *The Bells*. Text. The Bells, March 13. https://www.poets.org/poetsorg/poem/bells.

Pratha, Nimish, Natalie Avunjian, and Neil Cohn. 2016. Pow, Punch, Pika, and Chu: The Structure of Sound Effects in Genres of American Comics and Japanese Manga. *Multimodal Communication* 5 (October): 93–109. https://doi.org/10.1515/mc-2016-0017.

Pullman, Joost. 2001. Shaping Sounds in Comics. *The International Journal of Arts* 3 (1): 9–21.

Rakuten Kitazawa. n.d. Lambiek Comiclopedia. https://www.lambiek.net/artists/k/kitazawa_rakuten.htm. Accessed 8 May 2019.

Ramachandran, Vilayanur, and Edward Hubbard. 2001. Synaesthesia—A Window into Perception, Thought and Language. *Journal of Consciousness Studies* 8(12): 3–34.

Rhodes, Richard. 1995. Aural Images. In *Sound Symbolism*, ed. Leanne Hinton, Johanna Nichols, and John Ohala, 276–292. Cambridge: Cambridge University Press. https://doi.org/10.1017/CBO9780511751806.019.

Rohan, Olivia, Ryoko Sasamoto, and Rebecca Jackson. 2018. Argumentation, Relevance Theory and Persuasion: An Analysis of Onomatopoeia in Japanese Publications Using Manga Stylistics. *International Review of Pragmatics* 10 (2): 219–242.

Sapir, E. 1929. A Study in Phonetic Symbolism. *Journal of Experimental Psychology* 12 (3): 225–239. https://doi.org/10.1037/h0070931.

Sasamoto, Jun. 2000. Manga Hyogen Ni Okeru "Gion" No Hataraki [On the Function of "Gion" in Manga; Focusing on Its Role Word]. *Vi*.

Sasamoto, Ryoko. 2014. Impact Caption as a Highlighting Device: Attempts at Viewer Manipulation on TV. *Discourse, Context & Media* 6: 1–10.

Sasamoto, Ryoko, and Rebecca Jackson. 2016. Onomatopoeia—Showing-Word or Saying-Word? Relevance Theory, Lexis, and the Communication of Impressions. *Lingua* 175: 36–53.

Sasamoto, Ryoko, Minako O'Hagan, and Stephen Doherty. 2017. Telop, Affect, and Media Design: A Multimodal Analysis of Japanese TV Programs. *Television & New Media* 18 (5): 427–440. https://doi.org/10.1177/1527476416677099.

Sato, Eriko. 2017. Translanguaging in Translation: Evidence from Japanese Mimetics. *International Journal of Linguistics and Communication*. 5. https://doi.org/10.15640/ijlc.v5n1a2.

de Saussure, Ferdinand. 1959. *Course in General Linguistics*. New York: Philosophical Library.

Schieffelin, B.B., and E. Ochs. 1986. Language Socialization. *Annual Review of Anthropology* 15 (1): 163–191. https://doi.org/10.1146/annurev.an.15.100186.001115.

Schodt, Frederik L. 1983. *Manga! Manga!* New York: Kodansha America https://www.goodreads.com/work/best_book/948173-manga-manga-the-world-of-Japanese-comics.

———. 1996. *Dreamland Japan: Writings on Modern Manga*. Berkeley, CA: Stone Bridge Press.

Schwartz, Adam, and Eliane Rubinstein-Ávila. 2006. Understanding the Manga Hype: Uncovering the Multimodality of Comic-Book Literacies. *Journal of Adolescent & Adult Literacy* 50 (1): 40–49. https://doi.org/10.1598/JAAL.50.1.5.

Scott, Kate. 2017. Prosody, Procedures and Pragmatics. In *Semantics and Pragmatics: Drawing a Line*, ed. Ilse Depraetere and Raphael Salkie, 323–341. Berlin: Springer International. https://doi.org/10.1007/978-3-319-32247-6.

Scott, Kate, and Rebecca Jackson. Forthcoming. When Everything Stands Out, Nothing Does: Typography, Expectations and Procedures. In *Relevance Theory and Figuration*, ed. Piskorska Agnieszka. Amsterdam: John Benjamins Publishing.

Sell, Cathy, and Sarah Pasfield-Neofitou. 2015. Manga Studies: SFX Glossary—Onomatopoeia & Mimesis Translation. *The Manga Studies SFX Glossary*. http://mangastudies.com/sfx/?

Seto, Kenichi. 2003. *Kotoba Wa Aji Wo Koeru: Oishii Hyogen No Tankyu [Words Go Beyond Taste: Exploration of Tasty Expressions]*. Tokyo: Kaimeisha.

Shimizu, Isao. 1995. *Okamoto Ippei Manga Manbunshu [Collection of Manga and Manbun by Ippei Okamoto]*. Tokyo: Iwanami Shoten.

Shinohara, Kazuko, and Shigeto Kawahara. 2013. Otoshocho No Gengo Fuhensei: "ookisa" No Imeeji o Motoni [Linguistic Universality of Sound Symbolism: Based on the Images of Size]. In *OnomatopeKenkyū No Shatei: ChikazukuOto to Imi [Sound Symbolism and Mimetics: Rethinking the Relationship Between Sound and Meaning in Language]*, 43–57. Tokyo: Hitsuji Shobo.

Shinohara, Kazuko, and Ryoko Uno. 2013. *OnomatopeKenkyū No Shatei: ChikazukuOto to Imi [Sound Symbolism and Mimetics: Rethinking the Relationship Between Sound and Meaning in Language]*. Tokyo: Hitsuji Shobo.

Shiota, Eiko. 2003. Relevance Theory and Understanding Telops. *The Graduate School Review of the English Language and Literature* 31 (March): 63–91.

Shirane, Haruo, ed. 2002. *Early Modern Japanese Literature: An Anthology, 1600–1900*. New York: Columbia University Press.

Shitara, Kaoru. 2011. The Change of Telops on NHK Variety Shows. *Bulletin of Mukogawa Women's University, Humanities and Social Science* 59: 1–9.

Shohet, Merav. 2007. Narrating Anorexia: "Full" and "Struggling" Genres of Recovery. *Ethos* 35 (3): 344–382.

Shweder, Richard A., Robert A. LeVine, and Shweder Richard a, Social Science Research Council (U S.) Committee on Social and Affective Development During Childhood, and University of Cambridge. 1984. *Culture Theory: Essays on Mind, Self and Emotion*. Cambridge: Cambridge University Press.

Simner, Julia. 2012. Defining Synaesthesia. *British Journal of Psychology* 103 (1): 1–15. https://doi.org/10.1348/000712610X528305.

Simner, Julia, and Jamie Ward. 2006. The Taste of Words on the Tip of the Tongue. *Nature* 444 (7118): 438–438. https://doi.org/10.1038/444438a.

Simner, Julia, Louise Glover, and Alice Mowat. 2006a. Linguistic Determinants of Word Colouring in Grapheme-Colour Synaesthesia. *Cortex* 42 (2): 281–289. https://doi.org/10.1016/S0010-9452(08)70353-8.

Simner, Julia, Catherine Mulvenna, Noam Sagiv, Elias Tsakanikos, Sarah A. Witherby, Christine Fraser, Kirsten Scott, and Jamie Ward. 2006b. Synaesthesia: The Prevalence of Atypical Cross-Modal Experiences. *Perception* 35 (8): 1024–1033.

Skabelund, Aaron. 2011. *Empire of Dogs: Canines, Japan, and the Making of the Modern Imperial World*. 1st ed. Ithaca, NY: Cornell University Press.

Slater, Nigel. 2018. Nigel Slater's Crisp Aubergines with Chilli Sauce Recipe. *The Guardian*, February 20, sec. Food. https://www.theguardian.com/lifeandstyle/2018/feb/20/nigel-slater-crisp-aubergines-with-chilli-sauce-recipe.

Sneijder, Petra, and Hedwig F.M. Te Molder. 2005. Disputing Taste: Foods Pleasure as an Achievement in Interaction. *Appetite* 45 (1): 51–61. https://doi.org/10.1016/j.appet.2005.03.002.

Sohn, Ho-Min. 1994. *Korean*. Vol. 25. London and New York: Routledge.

Sontag, Susan. 1999. The Image World. In *Visual Culture: The Reader*, ed. Jessica Evans and Stuart Hall, 80–94. London and Thousand Oaks: Sage Publications in Association with the Open University.

Sperber, Dan, and Deirdre Wilson. 1986/1995. *Relevance: Communication and Cognition*, 2nd ed. Oxford: Blackwell.

———. 1998. The Mapping Between the Mental and the Public Lexicon. In *Thought and Language*, ed. Peter Carruthers and Jill Boucher, 184–200. Cambridge: Cambridge University Press.

———. 2002. Pragmatics, Modularity and Mind-Reading. *Mind & Language* 17 (1–2): 3–23. https://doi.org/10.1111/1468-0017.00186.

———. 2008. A Deflationary Account of Metaphors. In *The Cambridge Handbook of Metaphor and Thought*, 84–105. New York: Cambridge University Press. https://doi.org/10.1017/CBO9780511816802.007.

———. 2015. Beyond Speaker's Meaning. *Croatian Journal of Philosophy* 15 (2): 117–149.

Stevenson, John. 2005. *Yoshitoshi's Strange Tales*. Amsterdam: Hotei Publishing. https://www.worldcat.org/title/yoshitoshis-strange-tales/oclc/61666202.

Strauss, Susan. 2005. The Linguistic Aestheticization of Food: A Cross-Cultural Look at Food Commercials in Japan, Korea, and the United States. *Journal of Pragmatics, Focus-on Issue: Discourse and Metadiscourse* 37 (9): 1427–1455. https://doi.org/10.1016/j.pragma.2004.12.004.

Streeck, Jürgen, Charles Goodwin, and Curtis LeBaron. 2011. *Embodied Interaction: Language and Body in the Material World*. Cambridge: Cambridge University Press. https://books.google.ie/books?id=g7vqkiguszYC.

Sugimura, Yasushi. 2017. Nihongo No Onomatope "Hirihiri HiriQ, Hiriri", "Piripiri, PiriQ, Piriri", "Biribiri, BiriQ, Biriri" No Kijutsuteki Kenkyu [A Descriptive Study of Japanese Onomatopoeia "Hirihiri HiriQ, Hiriri", "Piripiri, PiriQ, Piriri", "Biribiri, BiriQ, Biriri"]. *Nagoya Daigaku Gengo Bunka Kenkyukai, Kotoba No Kagaku [Studia Linguistica]* 31: 111–130.

Szatrowski, Polly E. 2000a. Relation Between Gaze, Head Nodding and Aizuti "Back Channel" at a Japanese Company Meeting. *Annual Meeting of the Berkeley Linguistics Society* 26 (1): 283–294. https://doi.org/10.3765/bls.v26i1.1154.

———. 2000b. Kyoodoo hatuwa ni okeru sankasya no tatiba to gengo/higengo koodoo no kanren ni tuite [Relation Between Participants Status and Verbal/Nonverbal Behaviour in Co-construction]. *Nihongo Kagaku* 7: 44–69.

———. 2002a. Animeesyon No Sutoorii o Kataru Sai No Wadan to Tyuusin Hatuwa. *Hyoogen Kenkyuu* 76: 33–39.

———. 2002b. Nitibee in Okeru Animeesyon No Sutoorii No Katarikata to Higengo Koodoo No Sooi. In *Soogooteki Nihongo Kyooiku o Motomete*, ed. Osamu Mizutani and Deokbong Lee, 187–201. Tokyo: Kokusho Kankokai.

———. 2003. Gaze, Head Nodding and Aizuti "Back Channel Utterances" in Information Presenting Activities. *Japanese/Korean Linguistics* 11: 119–132.

———. 2004. *Hidden and Open Conflict in Japanese Conversational Interaction*. Tokyo: Kuroshio Shuppan. https://experts.umn.edu/en/publications/hidden-and-open-conflict-in-japanese-conversational-interaction-2.

———. 2005a. Danwa to Buntai—Kanzyoo Hyooka No Dooteki Na Katee Ni Tuit [Discourse and Style—On the Dynamic Process of Emotion/Evaluation]. In *Hyoogen to Buntai [Expression and Style]*, ed. Akira Nakamura, Mayumi Sakuma, Masaaki Nomura, and Chizuko Komiya, 468–480. Tokyo: Meiji Shoin. https://experts.umn.edu/en/publications/danwa-to-buntai-kanzyoo-hyooka-no-dooteki-na-katee-ni-tuite.

———. 2005b. Zyoohoo Syori, Soogo Sayoo, Danwa Koozoo Kara Mita Tooti to Higengo Koodoo to No Kankee. In *Katudoo to Site No Bun to Hatuwa [Sentences and Utterances as Activities]*, ed. Shuya Kushida, Toshiyuki Sadanobu, and Yasuharu Den, 159–208. Tokyo: Hitsuji Shobo.

———. 2008. Koogi Danwa Ni Okeru Sisee, Muimi Na Doosa, Miburi Ni Tuite. *Hyoogen Kenkyuu* 88: 24–37.

———. 2010a. Creating Involvement in a Large Japanese Lecture by Telling the Story of a Haiku. In *Storytelling Across Japanese Conversational Genre*, ed. Polly Szatrowski, 267–302. Amsterdam: John Benjamins Publishing. https://experts.umn.edu/en/publications/creating-involvement-in-a-large-japanese-lecture-by-telling-the-s.

———. 2010b. Koogi No Danwa No Higengo Koodoo. In *Koogi No Danwa No Hyoogen to [Expression and Understanding in University Lectures]*, ed. Mayumi Sakuma, 187–204. Tokyo: Kuroshio Shuppan. https://experts.umn.edu/en/publications/koogi-no-danwa-no-higengo-koodoo.

———., ed. 2010c. *Storytelling Across Japanese Conversational Genre*. Amsterdam: John Benjamins Publishing. https://experts.umn.edu/en/publications/storytelling-across-japanese-conversational-genre.

———. 2013. Tabemono o Hyooka-Suru Sai Ni Motiirareru "Kyakukanteki Hyoogen" to "Syukanteki Hyoogen" Ni Tuite [On the Use of "Subjective" and "Objective" Expressions for Food Assessment in Japanese]. *Kokuritu Kokugo Kenkyuuzyo Ronbunsyuu [NINJAL Research Papers]* 5: 95–120.

———., ed. 2014a. Introduction to Language and Food. In *Language and Food: Verbal and Nonverbal Experiences*, 3–31. Amsterdam: John Benjamins Publishing. https://benjamins.com/catalog/pbns.238.01sza.

———., ed. 2014b. *Language and Food: Verbal and Nonverbal Experiences*. Amsterdam: John Benjamins Publishing.

Tajima, Kaori. 2006. Onomatope (Giongo Gitaigo) Nit Suite. [On Onomatopoeia (Mimetics)]. *Kansai Gaikokugo Daigaku Ryuugakusei-Bekka Nihongo-Kyooiku Ronsyuu [Studies in Japanese Language Education at the International Center of Kansai Gaidai University]* 16: 193–205.

Takasaki, Midori. 2012. "Bimi" o Imi-Suru Go No Siyoo to Seesa- "Oisii" o Tyuusin Ni- [On Gender Differences and the Use of Words That Mean Bimi (Good Flavor) in Japanese: Focusing on Oisii (Delicious)]. *Otyanomizu Zyosi Daigaku "Zinbun Kagaku Kenkyuu" [Ochanomizu University "Studies in Arts and Culture"]* 8: 55–67.

Takatsuki, Yoshiteru. 2011. Manga Ni Okeru Hyogen Giho No Shinka—Nani Ga Manga o Bungei Ni Seichoo Saseta No Ka [Evolution of Expression

Techniques in Manga—What Made Manga a Literary Art?]. *The Bulletin of School of High-Technology for Human Welfare, Tokai University* 20: 53–75.

Takeuchi, K. 1998. Schôchôgo no honyaku o meguru futatsuno mondai [Two Issues on the Translation of Symbolic Expressions]. *Musashi daigaku jinbun-gakkai zasshi* 24 (4): 47–77.

Tamori, Ikuhiro, Hisao Kakehi, and Lawrence Clifford Schourup. 1996. *Dictionary of Iconic Expressions in Japanese*. Berlin and New York: Mouton de Gruyter. https://trove.nla.gov.au/version/26035751.

Tamori, Ikuhiro, and Lawrence Clifford Schourup. 1999. *Onomatope: Keitai-to Imi [Onomatopoeia: Form and Meaning]*. Tokyo: Kuroshio Shuppan.

Taruishi, Mako. 1992. *Atatakai Okurimono [Warm Gift]*. Tokyo: Fukuinkan.

Toratani, Kiyoko. 2005. A Cognitive Approach to Mimetic Aspect in Japanese. *Annual Meeting of the Berkeley Linguistics Society* 31 (1): 335–346. https://doi.org/10.3765/bls.v31i1.856.

———. 2013. Fukushiteki Onomatope No Tokushusei—Tagisei/Jishosei Kara No Kosatsu. In *Onomatope Kenkyu No Shatei: Chikazuku Oto to Imi*, ed. Kazuko Shinohara and Ryoko Uno, 85–99. Tokyo: Hitsuji Shobo.

———. 2018. Semantics and Morphosyntax of Mimetics. In The Cambridge Handbook of Japanese Linguistics, ed. Yoko Hasegawa, 202–221. Cambridge: Cambridge University Press. https://www.cambridge.org/core/books/cambridge-handbook-of-japanese-linguistics/semantics-and-morphosyntax-of-mimetics/25F0EC297C7BB5E257CE270C0CC5CAEB.

Toury, Gideon. 1995. *Descriptive Translation Studies – And Beyond*. Amsterdam and Philadelphia: John Benjamins.

———. 2004. *Los Estudios Descriptivos de Traducción y Más Allá. Metodología de La Investigación En Estudios de Traducción*. Madrid: Catedra. https://www.catedra.com/libro.php?codigo_comercial=130064.

Trask, R. Larry. 1993. *A Dictionary of Grammatical Terms in Linguistics*. London: Routledge. https://www.routledge.com/A-Dictionary-of-Grammatical-Terms-in-Linguistics-1st-Edition/Trask/p/book/9780415086288.

Tsujimura, Natsuko. 2001. Revisiting the Two-Dimensional Approach to Mimetics: Reply to Kita (1997). *Linguistics* 39 (2): 409–418. https://doi.org/10.1515/ling.2001.016.

———. 2006. Revisiting the Two-Dimensional Approach to Mimetics: Reply to Kita (1997). *Linguistics* 39 (2): 409–418. https://doi.org/10.1515/ling.2001.016.

Uchida, Yuzu, Junpei Watanabe, Jun Yoneyama, and Kenji Araki. 2011. Affect Analysis of Onomatopoeia Sentences Extracted from Blog Entries. In *17th Association for Natural Language Processing Conference*, 332–335.

Voeltz, F.K. Erhard, and Christa Kilian-Hatz, eds. 2001. *Ideophones*. Amsterdam: John Benjamins Publishing. https://benjamins.com/catalog/tsl.44.

Ward, Jamie, and Julia Simner. 2003. Lexical-Gustatory Synaesthesia: Linguistic and Conceptual Factors. *Cognition* 89 (3): 237–261.

Ward, Jamie, Julia Simner, and Vivian Auyeung. 2005. A Comparison of Lexical-Gustatory and Grapheme-Colour Synaesthesia. *Cognitive Neuropsychology* 22 (1): 28–41. https://doi.org/10.1080/02643290442000022.

Ward, Jamie, Brett Huckstep, and Elias Tsakanikos. 2006. Sound-Colour Synaesthesia: To What Extent Does It Use Cross-Modal Mechanisms Common to Us All? *Cortex* 42 (2): 264–280. https://doi.org/10.1016/S0010-9452(08)70352-6.

Ward, Jamie, Daisy Thompson-Lake, Roxanne Ely, and Flora Kaminski. 2008. Synaesthesia, Creativity and Art: What Is the Link? *British Journal of Psychology* 99 (1): 127–141. https://doi.org/10.1348/000712607X204164.

Watanabe, Chiemi, and Satoshi Nakamura. 2012. *Onomatoperori: Search Engine for Cooking Recipes by Impression of Their Tastes or Textures Using Japanese Onomatopoeia*. Kobe.

Westport Now Media. 2014. Big Lionel Has Found A Home—WestportNow.Com—Westport, Connecticut. *Westport Now*, September 9. https://westportnow.com/index.php?/v2_5/comments/meet_lionel/.

Wharton, Tim. 2003a. Interjections, Language, and the "Showing/Saying" Continuum. *Pragmatics and Cognition* 11 (1): 39–91.

———. 2003b. Natural Pragmatics and Natural Codes. *Mind & Language* 18 (5): 447–477. https://doi.org/10.1111/1468-0017.00237.

———. 2009. *Pragmatics and Non-Verbal Communication*. Cambridge: Cambridge University Press. https://doi.org/10.1017/CBO9780511635649.

———. 2010. Recipes: Beyond the Words. *Gastronomica: The Journal of Food and Culture* 10 (4): 67–73.

Wierzbicka, Anna. 1992. The Semantics of Interjection. *Journal of Pragmatics* 18 (2–3): 159–192. https://doi.org/10.1016/0378-2166(92)90050-L.

Wilkins, David P. 1992. Interjections as Deictics. *Journal of Pragmatics* 18 (2): 119–158. https://doi.org/10.1016/0378-2166(92)90049-H.

Wilson, Deirdre. 2003. Relevance and Lexical Pragmatics. *Italian Journal of Linguistics* 15 (January): 273–291.

———. 2004. Relevance and Word Meaning: The Past, Present and Future of Lexical Pragmatics. *Modern Foreign Languages (Quarterly)* 27: 1–13.

———. 2011. Parallels and Differences in the Treatment of Metaphor in Relevance Theory and Cognitive Linguistics. *Intercultural Pragmatics* 8 (2): 177–196. https://doi.org/10.1515/iprg.2011.009.

Wilson, Deirdre, and Robyn Carston. 2006. Metaphor, Relevance and the Emergent Property. *Mind & Language* 21 (3): 404–433.

———. 2007. A Unitary Approach to Lexical Pragmatics: Relevance, Inference and Ad Hoc Concepts. In *Pragmatics*, ed. N. Burton-Roberts, 230–259. Basingstoke: Palgrave.

Wilson, Deirdre, and Dan Sperber. 1993. Linguistic Form and Relevance. *Lingua* 90 (1): 1–25. https://doi.org/10.1016/0024-3841(93)90058-5.

———. 2002. Relevance Theory. *UCL Working Paper in Linguistics* 14: 249–290.

———. 2012. *Meaning and Relevance*. Cambridge: Cambridge University Press. https://www.cambridge.org/core/books/meaning-and-relevance/500D9D857CB3274BD23AF6FB381133B7.

Wilson, Deirdre, and Tim Wharton. 2006. Relevance and Prosody. *Journal of Pragmatics, Special Issue: Prosody and Pragmatics* 38 (10): 1559–1579. https://doi.org/10.1016/j.pragma.2005.04.012.

Wrembel, Magdalena. 2010. Sound Symbolism in Foreign Language Phonological Acquisition. *Research in Language* 8 (1): 1–14. https://doi.org/10.2478/v10015-010-0013-6.

Yoshida, Hanako. 2004. *Iconicity in Language Learning: The Role of Mimetics in Word Learning Tasks*. PhD Dissertation, Indiana University.

Yoshida, Hanako, and Linda B. Smith. 2003. Sound Symbolism and Early Word Learning in Two Languages. *Proceedings of the Annual Meeting of the Cognitive Science Society* 25 (25): 1287–1292.

Yoshikawa, Seiji, Shinya Nishimaru, Toyohisa Tashiro, and Masaki Yoshida. 1970. Collection and Classification of Words for Description of Food Texture III: Classification by Multivariate Analysis. *Journal of Texture Studies* 1 (November): 452–463. https://doi.org/10.1111/j.1745-4603.1970.tb00744.x.

Yus, Francisco. 2009. Visual Metaphor Versus Verbal Metaphor: A Unified Account. In *Multimodal Metaphor*, ed. C. Forceville and E. Uriós-Aparisi, 145–172. Berlín: Mouton de Gruyter.

Yus, Francisco, Charles Forceville, and Eduardo Uriós-Aparisi. 2009. Visual Metaphor versus Verbal Metaphor: A Unified Account. In *Multimodal Metaphor*, 145–172. Berlin: Mouton de Gruyter.

Zwicky, Arnold M., and Geoffrey K. Pullum. 1987. Plain Morphology and Expressive Morphology. *Annual Meeting of the Berkeley Linguistics Society* 7 (September): 330–340. https://doi.org/10.3765/bls.v13i0.1817.

Zwicky, Ann D., and Arnold M. Zwicky. 1980. America's National Dish: The Style of Restaurant Menus. *American Speech* 55 (2): 83. https://doi.org/10.2307/3050498.

Index[1]

A

B

C

[1] Note: Page numbers followed by 'n' refer to notes.

R. Sasamoto, *Onomatopoeia and Relevance*, Palgrave Studies in Sound,
https://doi.org/10.1007/978-3-030-26318-8

Papier | Fördert gute Waldnutzung
FSC® C083411

Kolibri 360 GmbH
Ferdinand-Jühlke-Straße 7
99095 Erfurt, Deutschland
produktsicherheit@kolibri360.de